T0143395

Recent Wireless Power Transfer Technologies via Radio Waves

RIVER PUBLISHERS SERIES IN COMMUNICATIONS

Series Editors

ABBAS JAMALIPOUR
The University of Sydney
Australia

MARINA RUGGIERI
University of Rome Tor Vergata
Italy

JUNSHAN ZHANG
Arizona State University
USA

Indexing: All books published in this series are submitted to the Web of Science Book Citation Index (BkCI), to CrossRef and to Google Scholar.

The "River Publishers Series in Communications" is a series of comprehensive academic and professional books which focus on communication and network systems. Topics range from the theory and use of systems involving all terminals, computers, and information processors to wired and wireless networks and network layouts, protocols, architectures, and implementations. Also covered are developments stemming from new market demands in systems, products, and technologies such as personal communications services, multimedia systems, enterprise networks, and optical communications.

The series includes research monographs, edited volumes, handbooks and textbooks, providing professionals, researchers, educators, and advanced students in the field with an invaluable insight into the latest research and developments.

For a list of other books in this series, visit www.riverpublishers.com

Recent Wireless Power Transfer Technologies via Radio Waves

Editor

Naoki Shinohara

Kyoto University
Japan

River Publishers

Published, sold and distributed by:
River Publishers
Alsbjergvej 10
9260 Gistrup
Denmark

River Publishers
Lange Geer 44
2611 PW Delft
The Netherlands

Tel.: +45369953197
www.riverpublishers.com

ISBN: 978-87-93609-24-2 (Hardback)
      978-87-93609-23-5 (Ebook)

Contents

PART II: Applications

PART III: Coexistence of WPT

**10 Human Safety on Electromagnetic Fields – The International
     Health Assessment**   **257**

Junji Miyakoshi

11 Coexistence of WPT and Wireless LAN in a 2.4-GHz Band   **277**

Koji Yamamoto and Shota Yamashita

Preface

Imagine a future where electricity can be supplied without any batteries or wires. In such a futuristic society, people will forget that electricity plays an important role in daily life. Mobile phones will last all day without any battery shortage. Trillion-sensor devices will collect big data ubiquitously and constantly, allowing novel insights into social problems both small and large. This future is not a dream. It is not science fiction. I am describing the near future, a society revolutionized by wireless power transfer (WPT) technology. WPT technology is based on radio-wave and wireless technologies that are primarily applied for wireless communication and remote sensing. This book describes the latest WPT technologies and cutting-edge WPT applications. I hope that you will share an electrified future without batteries or wires after reading this book.

Naoki Shinohara

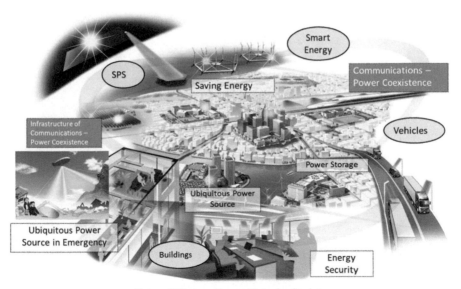

Future Wireless Power Transfer Society

List of Contributors

Aaron N. Parks, *Department of Electrical Engineering, University of Washington, USA*

Daniel Belo, *Instituto de Telecomunicações, Departamento de Electrónica, Telecomunicações e Informática, Universidade de Aveiro, Aveiro, Portugal*

Federico Alimenti, *Department of Engineering, University of Perugia, Italy*

Giulia Orecchini, *Department of Engineering, University of Perugia, Italy*

Joshua R. Smith, *Paul G Allen School of Computer Science and Engineering, and Department of Electrical Engineering, University of Washington, USA*

Junji Miyakoshi, *Kyoto University, Japan*

Ke Wu, *École Polytechnique de Montréal, Canada*

Koji Yamamoto, *Graduate School of Informatics, Kyoto University, Japan*

Luca Roselli, *Department of Engineering, University of Perugia, Italy*

Naoki Shinohara, *Research Institute for Sustainable Humanosphere, Kyoto University, Japan*

Nuno Borges Carvalho, *Instituto de Telecomunicações, Departamento de Electrónica, Telecomunicações e Informática, Universidade de Aveiro, Portugal*

Paolo Mezzanotte, *Department of Engineering, University of Perugia, Italy*

Ricardo Correia, *Instituto de Telecomunicações, Departamento de Electrónica, Telecomunicaçõeses e Informática, Universidade de Aveiro, Portugal*

Stefania Bonafoni, *Department of Engineering, University of Perugia, Italy*

Saman Naderiparizi, *Department of Electrical Engineering, University of Washington, USA*

Shota Yamashita, *Graduate School of Informatics, Kyoto University, Japan*

Simon Hemour, *Ecole Polytechnique de Montreal, Canada*

Tomohiko Mitani, *Kyoto University, Japan*

Valentina Palazzi, *Department of Engineering, University of Perugia, Italy*

Xiaoqiang Gu, *Ecole Polytechnique de Montreal, Canada*

Zerina Kapetanovic, *Department of Electrical Engineering, University of Washington, USA*

Zoya Popovic, *University of Colorado, Boulder, USA*

List of Figures

List of Tables

List of Abbreviations

ADC	Analog to Digital Converter
AM	Amplitude Modulated
AP	Access Point
ASK	Amplitude Shift Keying
CFA	Cross-Field Amplifier
CMOS	Complementary MOS
CNT	Carbon NanoTube
COMET	Compact Microwave Energy Transmitter
CSMA/CA	Carrier Sense Multiple Access with Collision Avoidance
dc	Direct Current
DNA	Deoxyribonucleic Acid
DR	Data Receiver
DT	Data Transmitter
DTV	Digital TV
EAS	European Space Agency
EHC	Environmental Health Criteria
EHS	Electromagnetic Hypersensitivity
ELF	Electromagnetic Field
EM	ElectroMagnetic
EMG	ElectroMyoGraphy
EPC C1G1	Electronic Product Code Class 1 Generation 2
EPC	Electronic Product Code
ES	Energy Source
ETHER	Energy Transmission Toward High Altitude Long Endurance Airship Experiment
FA	Frame Analyzer
FM	Frequency Modulated
FSK	Frequency-Shift-Keying
GRETA	GREen TAgs
GSM	Global System for Mobile communication
HSP	Heat-Shock Protein

I/Q	In phase/Quadrature
IARC	International Agency For Research On Cancer
ICT	Information and Communication Technologies
IoS	Internet of Space
IoT	Internet of Things
IR-UWB	Impulse-Radio Ultra-Wideband
ISM	Industrial Scientific and Medical
LEO	Low Earth Orbit
MAC	Media Access Control
MINIX	Microwave Ionosphere Nonlinear Interaction Experiment
MPT	Microwave Power Transfer
NASA	National AeroSpace Agency
NIH	National Institutes of Health
NTP	National Toxicological Program
OOK	On Off Keying
PACM	Phase-And-Amplitude-Controlled Magnetron
PC	Personal Computer
PCM	Phase-Locked Magnetron
PHY	Physical
PID	Proportional-Integral-Differential
PIR	Passive Infrared
PLL	Phase Loop Locked
PSK	Phase-Shift-Keying
PVPCM	Power-Variable Phase-Controlled Magnetron
RF	Radio Frequency
RFID	Radio Frequency Identification
SAR	Synthetic Aperture Radar
SCENIHR	Scientific Committee on Emerging And Newly Identified Health Risks
SHARP	Stationary High Altitude Relay Platform
SoC	System on Chip
SPORTS	Space Power Radio Transmission Systems
SPS	Solar Power Satellite
SWCNT	Single Wall Carbon NanoTube
TDMA	Time Domain Multiple Access
TWT	Traveling Wave Tube
UDP	User Datagram Protocol
UHF	Ultra High Frequency
UMTS	Universal Mobile Communication System

USA	United State of America
UV	Ultraviolet
UWB	Ultra Wide Band
VCO	Voltage Controlled Oscillator
WHO	World Health Organization
WISP	Wireless Identification And Sensing Platform
WLAN	Wireless Local Area Network
WPT	Wireless Power Transfer
WSN	Wireless Sensor Network

1

Introduction

Naoki Shinohara

Research Institute for Sustainable Humanosphere, Kyoto University, Japan

Abstract

In this chapter, the history of wireless power transfer (WPT) from the 19th century to the 21st century is described. WPT is a 21st century wireless technology, but it is based on very old and basic radio-wave technology from the 19th century. The technological underpinnings of WPT are summarized briefly in this chapter.

1.1 Introduction – History of Wireless Power Transfer

Wireless power transfer (WPT) technology has attracted considerable attention in the early 21st century. We already use wireless technology based on radio waves for broadcasting, information transmission, and remote sensing. We cannot imagine life without wireless technology. However, radio waves by themselves are not important for wireless technologies, but they are important information carriers.

Moreover, radio waves are used in energy-related applications, for example, microwave heating application by means of a microwave oven. After J. H. Poynting first derived the Poynting vector in 1884, which indicated that radio waves themselves have power, we have been using radio waves to provide heat. J. C. Maxwell derived Maxwell's equations in 1864. Maxwell's equations can explain not only radio waves but also light and electricity, which differ in terms of frequency alone. This indicates that we can use radio waves as electricity. This is WPT. Therefore, toward the end of the 19th century, N. Tesla carried out the first WPT experiment in Colorado Springs by using 150 kHz, 300 kW radio waves (Figure 1.1) [1].

Figure 1.1   Tower for WPT Experiment by N. Tesla (1899).

However, he failed because 150-kHz radio waves were too weak to supply adequate wireless power to users. Meanwhile, M. Hutin and M. Leblang patented an inductive wireless charger based on a 3 kHz magnetic field for electric vehicles [2]. Toward the end of the 19th century, electric vehicles were in use in Europe. It is well-known that G. Marconi successfully conducted the first wireless communication experiment in the same period. The end of the 19th century revolutionized wireless technology not only in terms of wireless communication but also wireless power.

In 1926, in Japan, an interesting WPT experiment was carried out by H. Yagi and S. Uda, the inventors of the Yagi-Uda antenna. They placed non-feed parasitic elements between a transmitting antenna and a receiving antenna with a frequency was 68 MHz to transmit wireless power. This device was named "wave canal," and it was similar to the Yagi-Uda antenna. They successfully received approximately 200 mW by transmitting 2–3 W of electric power [3].

WPT through radio waves was forgotten after the failure of Tesla's experiment, with the exception of Yagi-Uda's experiment. Instead, of the application of radio waves for power transmission, they were used for wireless communication and broadcasting. The world has changed because of the use of radio wave applications. According to Shannon's theorem, for increasing the amount of information transmitted by radio waves, their frequency must be increased. After World War II, microwaves were used in conjunction with vacuum tube technology, for example, Klystron, magnetron and TWT (Traveling Wave Tube). Antenna gain can be increased by increasing frequency. This means that we can concentrate wireless power at a target by using microwaves. W. C. Brown started a WPT experiment involving microwaves generated using a magnetron. He first developed a rectenna, a rectifying antenna, operating in the 2.45 GHz band. The rectenna always is applied for the WPT through microwave as a receiver and RF-dc converter. By using the developed rectenna and a magnetron, he developed the first microwave-driven drone (helicopter) in 1964 and 1968 in the US [4] (Figure 1.2). The flying drone received approximately 270 W of rectified 2.45 GHz microwaves. He and R. Dickinson succeeded in a 1-mile WPT field experiment in Goldstone in the US in 1975 (Figure 1.3). In this experiment, they used a 26-m-diameter parabolic antenna and a 450 kW Klystron as the transmitter. This is the biggest WPT experiment in the world. W. C. Brown

Figure 1.2   Microwave-assisted Drone Experiment by Brown (1964, 68).

Figure 1.3   450 kW WPT over 1 mile by Brown and Dickinson (1975).

also conducted a WPT experiment in the laboratory. In this experiment, total dc–dc efficiency of reached 54% was achieved by using 2.45 GHz microwaves in conjunction with a horn antenna and a magnetron. This is the highest efficiency achieved in any microwave-based WPT experiment worldwide. Based on the success of the microwave-based WPT experiments, a Solar Power Satellite (SPS) concept was proposed in 1968 [5]. The SPS is a future solar power station in space (36,000 km above the Earth), and the power generated by this station will be transmitted wirelessly to ground users. The microwave-based WPT system was too large from the viewpoint of commercial WPT applications. However, the SPS project has supported research and development of microwave-based WPT since the 1970s.

Brown's experiments and the SPS project are classified as "beam-type WPT" or "narrow-beam WPT." Instead of a wire, microwaves are concentrated in one receiver, and the transmission efficiency should be 100%. The efficiency of WPT can be calculated using Maxwell's equations and Friis transmission equation, and higher frequencies are required to increase the efficiency. Accordingly, Tesla's WPT experiment conducted using 150 kHz waves failed, while Brown's experiment conducted using 2.45 GHz waves succeeded. However, even when using microwaves, the existing efficiency levels remain inadequate, and system size remains larger than that suitable for commercial applications. Therefore, after the 1980s, R&D efforts on WPT have focused on simultaneous wireless communication and WPT by using radio waves, and one of the results of these efforts is RF-ID at 2.45 GHz and 920 MHz. Wireless power is distributed to multiple users, and the efficiency of this system is very low, similar to that of the wireless communication system. The WPT system is named 'Ubiquitous-type WPT' or 'wide-beam WPT'. Another WPT system for commercial applications is inductive WPT through LF magnetic field, where the distance between the transmitter and the receiver is almost zero. In the 1980s, commercial wireless chargers were produced for devices such as shavers, electric toothbrushes, and wireless telephones. Near Field Communication system, for example, IC card for traffic and banks, have been widely applied worldwide after the 1990s. There are many WPT applications around us, but we do not notice them. RF-ID is mostly a wireless communication system. NFC is considered a contact-powered system, not a wireless system. The hidden commercial WPT applications are the fruits of a few large R&D projects on inductive WPT chargers for electric vehicles, and these projects are based in California, US, France, and Germany after the 1980s. R&D on beam-type WPT using microwaves was also performed in universities worldwide [6]. New WPT technologies will be developed in the in 21st century as a result of these efforts.

In 2006, a research group based out of Massachusetts Institute of Technology (MIT) proposed a concept for inductive WPT called resonance coupling WPT [7]. They used resonance to increase the Q factor of an inductive coil. By using the developed high-Q coil, they achieved WPT by using 10 MHz waves over a distance of 2 m. One of the merits of inductive WPT is low cost and high efficiency. However, one of the demerits of inductive WPT is the short distance between the transmitter and the receiver. Resonance coupling WPT remedies this distance-related demerit while retaining its merits of low cost and high efficiency. After the development of resonance coupling WPT, the world finds the WPT technologies and acetate the R&D and commercialization of the WPT [8]. Now, a few large WPT consortiums involving many companies are promoting WPT chargers for mobile phones. Wireless chargers for parked electric vehicles (EV) and moving EVs are being developed worldwide, and a standard for such chargers is being discussed. These commercialized WPT applications are based on inductive and resonance coupling WPT, and they use frequencies around and below 100 kHz. In the US, venture companies have been established to promote microwave-based wireless chargers for mobile phones. Owing to these efforts, in the near future, battery-life-related anxiety will be eliminated, and wires will not be required for charging devices.

1.2 Wireless Power Transfer Technologies

Maxwell's equations can explain not only radio waves but also light and electricity, which differ in terms of frequency alone. Electricity is usually used in the form of Direct Current (DC) or 50/60 Hz AC. Given that a magnetic field and an electric field from a high-frequency current, we use it in the kHz-MHz order frequency. The magnetic field or the electric field can propagate only in near field which is enough shorter than wave length. Radio waves are mainly used over the MHz and GHz frequencies. Radio waves can be propagated over very far fields. We need a frequency converter from/to electricity to/from magnetic field or radio waves.

Therefore, a WPT system can be divided roughly into three parts: a frequency conversion circuit from electricity to high frequency or high frequency generator at a transmitter, an antenna or a coil to transmit wireless power between the transmitter and the receiver, and a frequency converter from high frequency to electricity or a rectifying circuit (rectifier) for conversion to dc at the receiver (Figure 1.4).

In each part, high efficiency is required. The total efficiency of WPT is obtained by multiplying the dc-RF conversion efficiency, transmission efficiency (beam efficiency via radio wave), and RF-dc conversion efficiency. In this book, technologies and applications of WPT through radio waves, especially, through microwaves, are described. The dc-RF conversion circuit technology is described in Sections 1.1 and 1.2. Antenna technology is described in Section 1.3. RF-dc conversion technology is described in Section 1.4. In Chapter 3, many interesting applications of WPT through

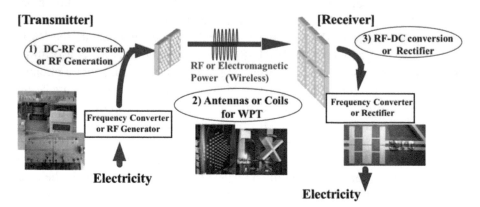

Figure 1.4 WPT System.

radio waves are introduced. Additionally, the safety and the EMC problem of WPT through radio waves are described in Chapter 4.

References

[1] Tesla, N. (1904). The transmission of electric energy without wires. *The thirteenth Anniversary Number of the Electrical World and Engineer* (Kempton, IL; Adventure Ultimate Press).

[2] Hutin, M., and Leblang, M. (1894). Transformer System for Elegtrig Rail-Ways. Patent US527857A.

[3] Yagi, H., and Uda, S. (1926). On the feasibility of power transmission by electric waves. *Proc. Pan Pac. Sci. Congr.* 2, 1307–1313.

[4] Brown, W. C. (1984). The history of power transmission by radio waves. *IEEE Trans. Microw. Theory Techn.* 32, 1230–1242.

[5] Glaser, P. E. (1968). Power from the sun; its future. *Science* 162, 857–886.

[6] Matsumoto, H. (2002). "Research on solar power station and microwave power transmission in Japan: review and perspectives," in *IEEE Microw. Magaz.* 3, 36–45.

[7] Kurs, A., Karalis, A., Moffatt, R., Joannopoulos, J. D., Fisher, P., and Soljačić, M. (2007). Wireless power transfer via strongly coupled magnetic resonances. *Science* 317, 83–86.

[8] Shinohara, N. (2011). Power without wires. *IEEE Microw. Magaz.* 12, S64–S73.

PART I

Technologies

2

Solid-State Circuits for Wireless Power Transfer

Zoya Popovic

University of Colorado, Boulder, USA

2.1 Introduction

Several methods for wireless power transfer have been demonstrated and published in the open literature, and are reviewed in Figure 2.1: (1) near-field reactive (inductive or capacitive) power transfer; (2) far-field directive beaming; (3) far-field non-directional low-power harvesting; and (4) wireless power transfer in an over-moded cavity. The method which is chosen is application dependent. For example, for high power transfer between the road and a stationary or moving vehicle, at 10–20 cm distance and for 10s of kW power transfer, near-field tuned reactive WPT in the kHz to MHz range is most reasonable. On the other hand, for harvesting very low ambient power densities of unknown sources for powering unattended sensors, far-field rectennas and arrays in the GHz range are useful, especially in environments with no light or vibration. In some space applications when cables are not a reasonable option, far-field power beaming has been proposed. On the other hand, if non-contact powering of multiple devices in a shielded environment is desired, an over-moded cavity can be used as the powering medium.

Solid-state circuits are used for both receivers and transmitters in WPT, and the devices and circuit architectures vary significantly depending on the power level and frequency dictated by the application. A general block diagram for the powering system is shown in Figure 2.2, but the specific dc-AC and AC-dc circuits, as well as their integration with the wireless far or near-field passive devices differs significantly based on power and frequency. The efficiency of the WPT system can be expressed as

$$\eta_{TOT} = \eta_T \cdot \eta_W \cdot \eta_R \qquad (2.1)$$

11

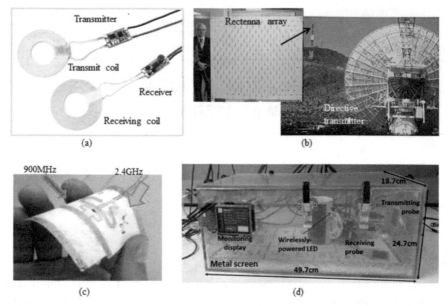

Figure 2.1   Fundamentally distinct types of WPT: (1) near-field reactive (inductive or capacitive) power transfer; (2) far-field directive beaming; (3) far-field non-directional low-power harvesting of multiple sources, e.g. in the cell-phone bands; and (4) wireless power transfer in an over-moded cavity.

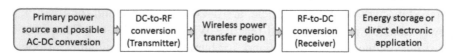

Figure 2.2   General power flow in a wireless powering system. The WPT region block in the middle is generally air or some other dielectric, and the transmitter and receiver can be in each other's near or far fields. The primary power source and energy storage blocks are not considered in detail in this chapter.

where η_T is the transmitter efficiency, η_W is the wireless transfer efficiency and η_R is the receiver efficiency. In this chapter, efficiency is used as one of the metrics for evaluating different WPT systems from Figure 2.1, starting from the low-power far-field harvesting applications.

2.2 Low-Power WP Harvesting

Power can be transferred by a propagating electromagnetic wave, radiated and received by antennas which do not load each other because they are in each other's far fields. In this case, the wireless transfer efficiency, η_W, is generally low compared to the near-field case, but does not necessarily depend critically on placement. The available incident power densities are on the order of micro-watts per cm^2 [1]. The power is radiated either from dedicated compliant low-power transmitter(s) which is in general narrowband or multi-frequency, or harvested from already available sources, which can be broadband and unknown to some degree. There is also ongoing work in modulated transmission, e.g. [2]. The position of the transmitters and receiving devices can vary and is not in general known precisely, requiring non-directional antennas and power reception circuits that can maintain efficiency over variable power levels which result from multipath. Because of the low available power densities, for any reasonable electronic application, the power needs to be collected over time and used when available. An example application is a low-duty cycle battery-less wireless sensor, e.g. [3]. In this case, the powering is performed independent of signal transmission, which differentiates it from RFID tags. A block diagram of such a sensor is shown in Figure 2.3.

In this low-power case, typically the transmitters are unknown sources designed for other applications, e.g. microwave ovens [1] or communication transmitters, [4, 5]. Therefore, in this case, the receiver circuit needs to be designed to enable rectification at variable and very low incident power density levels. Consider the block diagram of a wirelessly-powered wireless sensor (Figure 2.3), where an antenna integrated with a rectifier (referred to

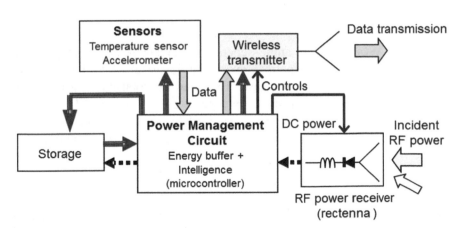

Figure 2.3 A block diagram of a wirelessly-powered wireless sensor with variable duty cycle of transmission. The power is collected in the far field of one or more ISM transmitters independently of data transmission. The sensor contains power management, sensor and transceiver circuitry.

as a "rectenna" in the literature) receives arbitrarily polarized radiation at one or more of the chosen frequencies at levels below often below 1 μW/cm². The dc output is managed by a digitally controlled power converter in such a way that it always presents close to an optimal dc load to the rectenna and transfers all input energy to the storage device, which provides power to the microcontroller, sensor and data transceiver. The sensor data is input to a commercial low-power wireless transceiver operating in, e.g. the 2.4 GHz ISM band. The data transmission is the most power-consuming task and is not done continuously, which is acceptable for most applications. If there is not enough stored energy, the data cannot be transmitted and there is danger of damaging the storage device. Therefore, the available rectified RF power and the available stored energy are monitored in a closed-loop system allowing for adaptive adjustment of the data transmission duty cycle.

The efficiency of the wireless power reception includes the efficiency of the integrated antenna-rectifier (rectenna), η_{RA}, and the converter efficiency, η_C, and can be written as [6]:

$$\eta_R = \eta_{RA} \cdot \eta_C = \frac{P_{R,dc}}{P_{RFinc}} \cdot \frac{P_{harv}}{P_{R,dc}} = \frac{P_{harv}}{P_{RFinc}} \qquad (2.2)$$

where $P_{R,dc}$ is the dc power output of the rectenna, P_{harv} is the dc power delivered to the storage element, and P_{RFinc} is the incident power on the rectenna of geometric area A_G given by

$$P_{RFinc} = S \cdot A_G, \qquad (2.3)$$

where $S = S(\theta, \phi)$ is the angle-dependent incident power density of one or more plane electromagnetic waves, and is assumed to be less than 100 μW/cm² in the harvesting case. To obtain the total incident power, in general the total power density is integrated over a sphere. The rectenna efficiency defined in this way is determined by measuring the dc voltage across a known load R_L and by calibrating the incident power density:

$$\eta_{RA} = \frac{P_{R,dc}}{P_{RFinc}} = \frac{V_{dc}^2}{R_L} \cdot \frac{1}{S \cdot A_G} \qquad (2.4)$$

where S is the power density at normal plane-wave incidence, assuming a single transmitter in the far field and an electrically small antenna with constant S across its aperture. Multiple transmitters can be taken into account by adding the dc powers resulting from a superposition of plane waves. Note that this is the most conservative definition, since the geometric area of the

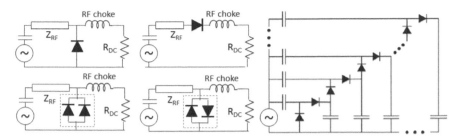

Figure 2.4 Single-ended diode rectifiers (series and shunt), parallel and anti-parallel dual diode rectifiers and Dickson charge pump (voltage multiplier) multi-diode rectifier.

antenna is always smaller than its effective area, and Equation (2.4) thus takes into account the aperture efficiency and losses of the antenna, the impedance mismatch between the antenna and the rectifier, and the losses in the rectifier circuit. The rectenna efficiency is a nonlinear function of incident power due to the nonlinear impedance of the rectifier, and the above quantities depend on frequency.

Most rectifiers for low power levels have been implemented with Schottky diodes, either single-ended (shunt or series), multiple diodes or in a Dickson pump (voltage multiplier) circuit used to increase the voltage output, Figure 2.4. A limited comparison of single and multiple-diode rectifiers, as well as a detailed nonlinear analysis, is given in, e.g. [7]. When the incident power is very limited, a single diode is the best option since turning each diode on requires some threshold voltage. The highest rectification efficiency is obtained when the diode rectifier is impedance matched to the antenna at the predicted power level, since the diode impedance varies with power level. A design and verification procedure that results in optimized rectenna efficiency as defined by Equation (2.4) is presented next, and results given for an example linearly-polarized patch antenna integrated with a single Schottky diode rectifier at 1.96 GHz. The method is then applied to a dual-polarized rectenna at 2.45 GHz, with an investigation of an optimized dc collection circuit design. The results of this section are used as the input to the power management circuit design.

The first step is rectifier characterization, since the antenna needs to be impedance matched to this nonlinear RF load. The impedance for optimal rectification is not the same as that for optimal reflection coefficient and needs to be characterized using nonlinear load-pull modeling or measurements, as described in Figure 2.5. In contrast to a network-analyzer characterization, the RF input impedance and the dc output load and input RF power are varied.

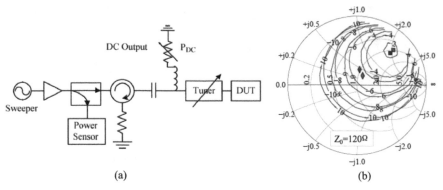

(a)   (b)

Figure 2.5   (a) Block diagram of modified load-pull measurement and simulation setup for nonlinear characterization of rectification efficiency of a shunt Schottky diode rectifier. (b) Example measured constant efficiency contours (blue) and simulated (red) for 0 dBm (1 mW) RF power incident on the diode, with $R_{dc} = 1050\ \Omega$ at 1.96 GHz. The square symbols indicate the impedances for best efficiency. The diamond symbols show best efficiency points for $R_L = 63\ \Omega$.

The output quantity is the dc power across the variable load. The optimal dc load is chosen to keep the voltage across the rectifying element between forward-bias threshold and breakdown.

The result of the load-pull measurement or analogous nonlinear harmonic-balance simulation are a set of constant dc power contours, which are best plotted on an impedance Smith chart as shown in Figure 2.5. In this plot, measured data using a commercial load-pull system are compared to harmonic-balance nonlinear simulations using a diode model in Keysight's ADS circuit simulator. Example contours are shown for a Skyworks SMS7630-79 diode at one power level, 0 dBm, which for an antenna that is 5 cm × 5 cm in size corresponds to $S = 40\ \mu W/cm^2$. The RF impedance for optimal efficiency can be found from this analysis, and varies over dc loads, frequency and between different diodes.

To complete the power receiving circuit, the antenna impedance is designed to match the range of optimal rectifier impedances over the power range of interest, and the design of the dc collection circuit. For the example given here, where the application is harvesting power from cell-phone basestations [8], the optimal impedance at 1.96 GHz (cell band) is chosen to be $137+j149\Omega$, based on nonlinear characterization. A linearly-polarized probe-fed patch antenna designed for 1.96 GHz is shown in Figure 2.6(a), while Figure 2.6(b) shows the back-side microstrip circuit that matches the feed-point impedance to the desired impedance. The matching procedure

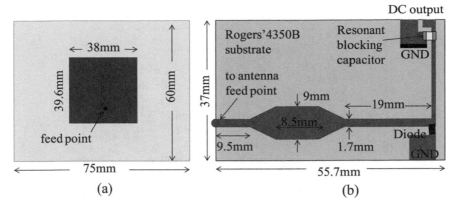

Figure 2.6 Layout of a 1.96 GHz linearly-polarized patch antenna with a coaxial probe feed connected to a microstrip matching and dc collection circuit through a via in the common ground plane. (a) Top substrate antenna side and (b) lower substrate circuit side. The two substrates share a ground plane.

is described in detail in [7]. The high-impedance (narrow) line moves the antenna impedance to the real axis, while the low-impedance line acts as a transformer. The additional 19-mm line brings the impedance finally to the diode impedance for optimal efficiency at the desired power level, from simulated and measured source-pull data. The dc collection circuit is designed to be an open circuit at the RF frequencies. The antenna and matching circuit are fabricated on Rogers 4350b, 0.762-mm thick substrate, and are simulated using Ansoft HFSS.

The characterization of the integrated rectifier-antenna is performed in the far field as shown in Figure 2.7. In order to accurately determine the incident power density, a calibrated horn antenna is first placed at the reference plane of the rectenna and power measured with a power meter. For such calibrated power densities in the range of 20–200 μW/cm^2, the power across a range of dc loads is measured for the rectenna placed at the reference plane as shown in Figure 2.7, from which the efficiency given by Equation (2.4) can be calculated and the resulting values are shown in Figure 2.8.

A wirelessly-powered sensor receives power radiated from a transmitter in the far field, and as a result of the variety of locations that sensors can be placed at, and the different orientation of the sensor relative to the power transmitter, it is important to characterize the performance of the integrated rectenna at different angles, in addition to different incident power densities. The dc pattern can be directly measured at the output of the rectenna, and is

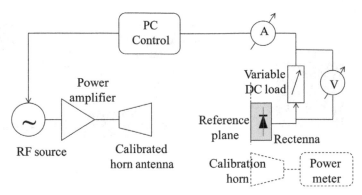

Figure 2.7   Block diagram of rectenna measurement setup. A digitally controlled dc load is used to determine experimentally the optimal load for a given input power at a given frequency. The rectenna is placed at the reference plane where the incident power density is calibrated through a known horn antenna.

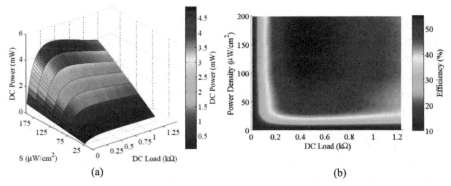

(a)   (b)

Figure 2.8   (a) Measured co-polarized rectified power for the rectenna shown in Figure 2.6 at the broadside for power densities from 25 to 200 μW/cm^2. Every intersection of the black grid lines and black curves is a measured data point. (b) Rectenna efficiency calculated from measured data in (a) and Equation (2.4). The load resistance is varied from 0 to 1250 Ω, showing that the best efficiency occurs for an optimal dc load around 460 Ω in this case.

the product of the antenna power pattern and the rectifier efficiency for the appropriate power density at each angle.

The efficiency drops rapidly at lower input powers, and this nonlinearity of the rectification process can be used to predict the difference between patterns of a rectenna compared to the antenna radiation pattern alone. An example comparison is showed in Figure 2.9, where the normalized

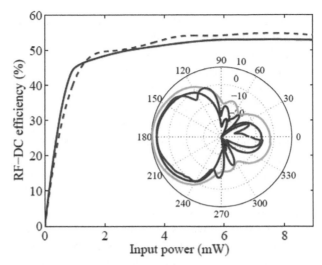

Figure 2.9   Measured rectenna efficiency (dashed red) and nonlinear diode rectifier simulated efficiency multiplied by the simulated antenna radiation efficiency (blue). Inset: Normalized measured co-polarized E-plane 50-Ω patch antenna pattern (green). Predicted dc pattern for a incident power density of 200 μW/cm^2 (blue) and measured rectenna pattern at 200 μW/cm^2 (red). The H-plane shows similar behaviour. The angle is in degrees and the power in dB.

RF antenna pattern is shown together with predicted and measured rectenna dc pattern, which is approximately 3 dB below the RF power pattern and in excellent agreement with the maximum measured efficiency at broadside of 54% for a power density of 200 μW/cm^2. The prediction of dc power patterns can be scaled to other power densities.

The rectenna output power needs to optimally charge a storage element. The purpose of the power management circuit is to act as a buffer between the rectenna power source and the energy storage device (Figure 2.3). To act as an ideal buffer in the harvesting application, the converter must perform three functions: (1) create at its input port the optimal impedance match to maximize the rectenna efficiency, over the full range of incident power densities; (2) transfer the harvested energy with minimal loss to the energy storage element over the full range of rectenna output voltages and energy storage charge states; and (3) monitor the energy storage and provide charge control and protection as appropriate for the energy storage used (battery or capacitor). Since the efficiency of the rectenna depends on the matching behavior of the converter, and the efficiency of the converter depends on the operating conditions of the rectenna and the energy storage device, it is best to co-design these blocks for the given application.

The first function of the converter is to maximize the rectenna efficiency by creating a converter input port that emulates the optimal load impedance of the rectenna [6, 8, 9]. The filter integrated in the rectenna creates a dc port and reduces the rectenna model from the perspective of the power converter to a Thevenin equivalent, and the rectenna output impedance reduces to an equivalent resistance. Thus, the optimal load to the rectenna is a dc resistance, apparent in the measurement results of Figure 2.8. The ideal converter, Figure 2.10, is modeled with an input port that emulates a resistor, R_{em}, and an output port that transfers all of the power from the input port to the energy storage device, shown as a battery model. The challenge in the low power harvesting application is to perform the behavior of Figure 2.10 with minimal control circuit overhead so that the control losses can be kept small when compared to the power being processed. This rules out many of the advanced control circuits and techniques commonly applied at higher power levels. A simple approach that requires minimal control overhead is to use a boost converter to provide the required step up from typical rectenna voltages of tens to hundreds of millivolts to typical battery voltages, from 2V to 4V. The key to achieving a good match to the rectenna is found in the timing control circuit and the resulting inductor current waveform, as detailed in [8]. Optimization of the power converter involves selection of the power stage components, power semiconductor devices, and control circuit design. At lower power levels the loss is dominated by control circuit losses and at higher power levels conduction losses dominate.

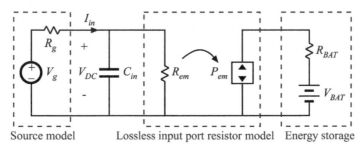

Source model   Lossless input port resistor model   Energy storage

Figure 2.10   Ideal lossless input port resistor model of the power converter, showing an input port that emulates a resistor, R_{em}, and an output port that transfers the power from the input port to the output energy storage, shown here as a battery.

One example developed in [8] is harvesting RF energy from cellular base station towers, where the output varies over at least two orders of magnitude in a 24 hour period and varies widely from one tower to another. In that design, an adaptive control circuit with online optimization was implemented using a Texas Instruments MSP430 microcontroller as the timing control circuit. The hardware results shown in [7, 8] are based on low-cost off-the-shelf components and demonstrated record efficiencies for power levels down to approximately 100 μW. Below this level, the quiescent losses of available control hardware become the limiting factor. With the significant interest and advancements in ultra low power wireless sensors, a range of custom solutions have been developed for power conversion at these low power levels, e.g. [9–11]. A power converter designed for both high efficiency operation below 100 μW and input port matching to the rectenna for energy harvesting is presented in [12]. The IC uses deep sub-threshold design techniques to achieve a nominal quiescent supply current of 200 nA. A photograph of the IC is shown in Figure 2.11 along with experimental results of the converter efficiency for both online optimization using an external MSP430 microcontroller and manual timing adjustment. This IC is connected to a 6 cm × 6 cm patch rectenna and a 2.5 V battery, with a positive output power at incident power densities below 2 μW/cm^2, converter efficiency above 70% down to

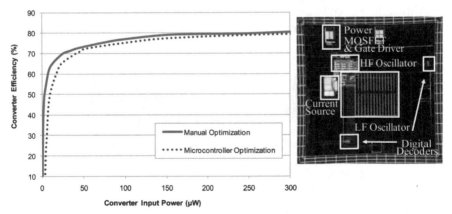

Figure 2.11 Measured converter efficiency with the custom energy harvesting IC with manual and microcontroller based optimization of control timing parameters. The dashed line is for the case of microcontroller optimization and solid line is manual. A photograph of the 2 mm × 2 mm 0.35 μm 5-V CMOS chip is shown on the right.

30 μW of input power, and a converter emulated resistance that varies less that $+/-$ 5% over the full operating range, resulting in a matching efficiency greater than 95%.

An example of system level integration with load management is given in [13], and a photograph of the integrated wireless sensor board is shown in Figure 2.12. The sensors arc operated in a low duty cycle burst mode and data only taken and transmitted at a low sampling rate, on the order of Hz. The load management algorithm changes the sampling rate and system mode based on the measured battery voltage (SOC) and RF input power. The variable sampling rate is used to regulate the battery SOC and achieve energy balance in the system.

Different applications might call for different rectifier and antenna topologies. Although an advantage of the patch antenna is that the circuitry can be placed behind the antenna ground and mounted on any object, dipole antennas with no ground plane can be used when less directional coverage is required, or dipoles with reflectors can be used for higher directivity [14]. Antennas can be fabricated on a variety of substrates, including flexible substrates with ink-jet printing, e.g. [5, 15]. Rectennas can also be arrayed at the dc output to obtain higher voltages or currents, e.g. [16–18]. Other circuit topologies, such as full-wave rectifiers or charge pumps can be employed for increased efficiency at higher input power levels.

Although we consider the entire powering system with a low-power FCC-compliant transmitter, multiple or modulated transmitters can also be

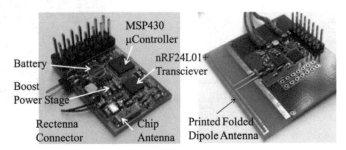

Figure 2.12 Photograph of power management circuit for a wireless sensor with power harvesting based on off-the-shelf components. The 2.1 cm × 1.7 cm circuit contains power management, sensor and transceiver circuitry. Data collected from the sensors is transmitted from a chip antenna (*left*) or printed-circuit-board folded dipole (*right*). The circuit is mounted on the ground plane behind an integrated patch antenna rectifier which provides the wireless power.

considered, or ambient energy can be harvested. Power scavenging is demonstrated in [4] from emissions from 470–570 MHz TV broadcast transmitters, and in [14] and [15] two common cellular/ISM bands around 900 MHz and 2.4 GHz. The dual-circularly polarized spiral array in [17] addresses 2–18 GHz broadband harvesting at very low incident power densities, Figure 2.13. In this work, 10,000 randomly generated two-tone signals at frequencies between 2 and 8 GHz with various power levels were incident on a dual-circularly polarized rectenna array, and in all cases measured increased dc power relative to the sum of the same two individual tones. This concept was extended to chaotic signal waveforms in [2] and a study with various peak-to-average ratio (PAR) signals is shown in [19].

There are a number of applications for low-duty cycle sensors where batteries are difficult or impossible to replace, e.g. sensors for health monitoring of patients, aircraft structural monitoring, sensors in hazardous environments, sensors for covert operations, etc. One interesting example is harvesting power of sidelobes from an altimeter radar antenna on a commercial aircraft for aircraft structural health monitoring [20]. The device needs to deliver at least 300 µJ of energy during a time span of 10 minutes to be able to energize a low power, low duty cycle sensor, resulting in an average power requirement of P = −33 dBm. Based on this average dc power that rectifier has to provide, a number of commercial diodes are considered and a source-pull simulation is performed in the NI/AWR Microwave Office harmonic balance simulator

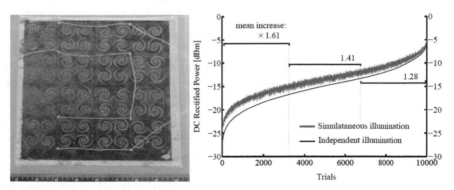

Figure 2.13 Arrays can be used for scalable energy scavenging. Comparison of total recitified power for independent and simultaneous dual-frequency (between 2 and 8 GHz) illumination with a variable multi-tone signal (power variation from 0.1 µW/cm² to 0.1 mW/cm²). The results of 10,000 trials are rank-ordered by dc power to illustrate the power dependent increase in rectification efficiency.

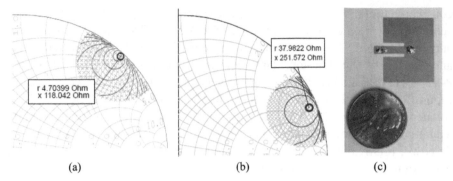

(a)   (b)   (c)

Figure 2.14   Source pull simulations showing the optimal RF source impedance at 4.3 GHz for maximum rectifying efficiency, for (a) Skyworks SMS7621-079 diode (Modelithics model) and (b) Skyworks SMS7630-061 diode (Spice model). Photograph of implemented rectenna (c).

using available nonlinear models. The optimum RF impedance presented to the diode for the highest rectifier efficiency is determined, and the devices that require the smallest input power selected: the Skyworks SMS7621-079 and SMS7630-061 GaAs Schottky diodes. Figure 2.14 shows the source pull simulation results for both devices. In all cases, the input power is set to obtain the required −33 dBm output power, and the dc load is swept to find the optimum in terms of rectification efficiency.

An integrated patch rectenna at 4.3 GHz with the dc collection point taken from the patch RF null is also shown in Figure 2.14. Measurements are carried out inside an anechoic chamber with calibrated power densities at the plane of the rectenna of 0.13 μW/cm^2 and 0.65 μW/cm^2, both within the lower range of the estimated power density in the harvesting environment, performed from measured altimeter antenna radiation patterns. Two values of capacitors are used as energy storage devices, 100 μF and 1 mF.

Figure 2.15(a) shows the open circuit voltages for all the prototypes, where measurements at the higher power level are only shown for the SMS7630-061 prototype. The open-circuit voltage is used to find the resonant frequency of the device and to calculate the stored energy in the capacitor, and the peak open circuit voltage of 57 mV with 0.16 μJ is measured at 0.13 μW/cm^2 with a 100 μF capacitor, while 218 mV with 23.8 μJ is measured at 0.65 μW/cm^2 with a 1 mF storage element. Open circuit voltage and energy are not sufficient to characterize the behavior of the devices since the stored energy must be collected in a fixed amount of time. For this

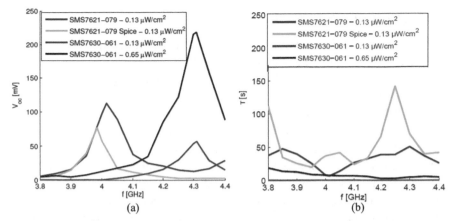

Figure 2.15 (a) Measured open-circuit voltage of the prototypes using the different diodes. (b) Time constant τ, measured as the time it takes the circuit to reach 63% of its open-circuit voltage.

reason, measurements of the time constant of the charging circuit were taken, Figure 2.15(b), where it is seen that prototype SMS7630-061 is considerably better, and can store the energy in shorter bursts than the other two prototypes. The average power delivered to the capacitor in one time constant is above −31.5 dBm in the 4.2–4.4 GHz band of interest.

The above examples illustrate the design procedure for low-incident power receivers in harvesting WPT systems. A large variety of rectennas for this application has been demonstrated to date, including dipoles, crossed dipoles, slots, patches, Yagi-Uda arrays, spirals, etc. Various rectifier topologies have also been demonstrated, with some examples overviewed in [21]. For energy harvesting, broad bandwidth or multi-band operation is essential. Example dual-frequency scavenging rectennas are shown in [5] and [15], while broadband arrays and elements are discussed for example in [22] and [23] and demonstrated over almost a decade with broad bandwidth and dual polarization in [17]. Approaches for ambient harvesting integrated devices is shown in a wristband rectenna at 2.45 GHz [24] and applied to sensors in [25]. The fundamental limit in receiving very low power levels is that diodes turn on at finite voltages. Although zero-bias Schottky diodes are the most common choice to date, some research into zero turn-on voltage diode rectifiers is presented in, e.g., [26] and [27], with the goal of ultra-low power rectification for harvesting applications.

2.3 Medium-Power WPT

Medium-power WPT in several watt ranges at lower frequencies is commercially available for personal devices and is typically done reactively at very short distances through a tightly-coupled transformer using standard circuits, e.g. [28]. The usual operating frequencies are around 100 kHz, although more recently ISM bands in the 6–27 MHz range have also been used. One of the reasons for increasing the frequency from the kilohertz to the megahertz range is to improve the tolerance to misalignment by eliminating ferrite cores. A higher frequency also increases the reflected resistance seen by the primary coil driver, therefore, power can be transferred at reduced current stresses. Similar to switched-mode power supplies, the additional benefits of operating at MHz ISM bands are reduced values of the passive components, increased power density, improved performance and transient response, and lower coupling with nearby objects and devices. A large body of work is published in this frequency range in the power electronics community, and some examples with efficient solid-state circuit design of inverters (transmitters) and rectifiers (receivers) are shown in [29–32].

In this chapter, the focus is on microwave WPT which is the more difficult frequency range for high efficiency implementations. At watt-level power levels, safety regulations become an issue for far-field GHz-frequency powering, and in this section an example of a shielded WPT system is shown to illustrate medium-power solid-state circuits for this type of applications. Figure 2.16 shows a heavily over-moded shielded metal waveguide cavity at 10 GHz [33] excited through patch probes by watt-level efficient power amplifiers. When multiple wirelessly powered devices are placed in the cavity, a statistical power density distribution results [34]. The power distribution becomes more uniform when more devices are charged (load stirring) and with mechanical or frequency stirring, as shown in the measured and simulated histograms in Figure 2.16, for 11 scatterers placed in the cavity, simulating devices under charge, and for 2000 possible measured relative device positions.

2.3.1 Medium-Power Microwave Transmitter Circuits

For 1–10W power levels, highly efficient power amplifiers have been demonstrated in both GaAs and GaN, with a good overview of the various approaches given in [35]. In the 2-GHz range, power added efficiencies above 85% have been shown with 250-nm GaN on SiC devices, e.g. [36], and at 10 GHz, efficiencies close to 70% have been shown in MMIC PAs, e.g. [37] and [38]. In these amplifiers, the transistor is driven into a highly

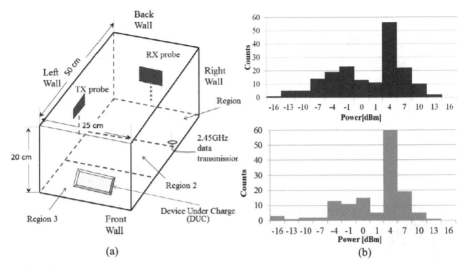

Figure 2.16 (a) An over-moded waveguide cavity can be used for powering multiple devices simultaneously. One such device under charge (DUC) is shown located in Region 3 of the cavity. The three regions are defined for measurement purposes. Transmit and receive 10-GHz patch probes are located on two walls of the cavity, which measures $8.3\lambda_0 \times 6.6\lambda_0 \times 16.6\lambda_0$, supporting a large number of modes. (b) Measured (top) and simulated (bottom) power distribution for 2000 spatial combinations of 11 DUTs positioned within the cavity, with a single 0.25W transmitter. The power is received with a patch probe in all cases.

nonlinear regime and generated harmonics are used to perform waveshaping of the current and voltage time-domain waveforms across the virtual drain (current source) of the device, minimizing the $v(t) \cdot i(t)$ product and thus maximizing efficiency. Although not appropriate for communications due to large signal distortion, these amplifiers are excellent candidates for powering, especially in the shielded approach in Figure 2.16. Some examples of PAs and measured CW properties are shown in Figure 2.17 for S and X-band GaN amplifiers.

2.3.2 Medium-Power Microwave Rectifier Circuits

In either far-field powering or in the over-moded cavity from Figure 2.16, the incident powering wave will in general reach the rectenna (power receiver) with a polarization that varies in time, depending on the radiating source and propagation environment. The dependence on alignment of the source to the receiver should be minimized for reduced positioning sensitivity. A dual-linearly (or circularly) polarized rectenna, which rectifies the power contained

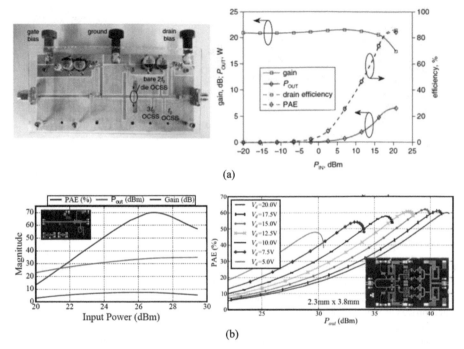

(a)

(b)

Figure 2.17   (a) Hybrid 2-GHz class-F^{-1} PA using a Qorvo GaN on SiC unpackaged transistor, with measured CW performance showing PAE > 85% with 7 W output power. (b) Monolithic single-stage 10-GHz GaN PA with 5 W output power and PAE ∼ 70% at peak power and G < 10 dB. (c) Monolithic two-stage power combined 12-W peak power PA with G > 20 dB and peak efficiency above 55% over a range of output powers obtained by varying the supply voltage.

in orthogonally-polarized waves independently and adds the dc output, will on average receive the most power with the least variation over time and with increased overall efficiency over a linearly-polarized rectenna [39]. This property has been recognized by many of the rectenna integrated rectifier antenna designs in the literature. In [40–43] dual linear square patches are used with a rectifier for each of the polarizations where the dc outputs are combined at a single node. In [45] a circular polarization patch has one port is terminated in 50 Ω while the other is used for power reception, so the rectifier antenna needs to be configured according to the incoming wave. A crossed dipole is used in [46], with dc outputs connected across the same dc load.

   The design of a dual-polarized patch rectenna for medium power level transmitters is shown here as an example. In this case, due to increased power density levels in the mW/cm^2 range (compared to μW/cm^2 in the previous section), harmonics are generated at the output of the rectifier and can be used to improve rectification efficiency. In many of the reported microwave rectifiers, filtering harmonics at both the input and output has been investigated, e.g., [47], mainly to reduce re-radiated harmonic power. However, the impact of harmonic terminations on the wave shaping and resulting RF-dc conversion efficiency can also be effectively used, in analogy to PA harmonically-terminated classes of operation [48]. In a rectifier, the nonlinear rectifying element generates currents and voltages at the harmonics of the input frequency, and the efficiency of the rectifier can be increased by properly terminating the harmonic frequencies at the diode reference plan, thereby reducing the overlap with current and voltage time-domain waveforms. This is illustrated theoretically in Figure 2.18(b) for a reduced-conduction angle rectifier (class C), where all harmonics are ideally short-circuited. The predicted time-domain waveforms are shown in Figure 2.18(c) for a Skyworks SMS7630 Schottky diode nonlinear model with a harmonic-balance simulation and with 5 terminated harmonics corresponding to the experimental rectifier circuit.

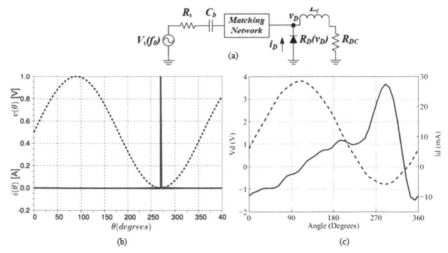

Figure 2.18   (a) Single-ended microwave rectifier circuit diagram. An ideal blocking capacitor C_b provides dc isolation between the microwave source and rectifying element. An ideal choke inductor L_c isolates the dc load R_{dc} from RF power. (b) Ideal normalized voltage (dashed) and current (solid) waveforms for reduced conduction angle half-wave rectifier (class-C with infinite harmonics shorted). (c) Harmonic balance simulation of current and voltage waveforms for five harmonic terminations.

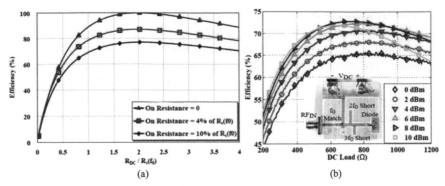

(a)   (b)

Figure 2.19   (a) Simulated efficiency $R_{dc}/R_S(f_0)$ for varying rectifier on-resistance for an infinite number of harmonics and the waveform in Figure 2.18(b). (b) Measured efficiency of rectifier circuit.

Since the current only conducts for a very short time, ideally the only impact on efficiency will be due to the on-resistance of the diode and Figure 2.19(a) shows calculated efficiencies from a Fourier expansion analysis of the class-C waveforms and the rectifier idealized circuit model. The Skyworks SMS7630 Schottky diode in the SC-79 package was source-pulled at 2.45 GHz with 0–10 dBm available input power for various dc loads in order to identify the combination of input power, fundamental load and dc load resulting in highest efficiency, as in Figure 2.5. The on-resistance of the SMS7630 is 20 Ω with the optimal dc load being found as 1080 Ω, therefore R_{ON} is approximately 2% of R_{dc} and from Figure 2.3, the ideal upper limit on peak efficiency is 87%. In the design of the rectifier circuit, 5 harmonics are taken into account, and the measured efficiency of the class-C rectifier is shown in Figure 2.19(b).

A dual-polarized 40-mm square patch antenna is designed at 2.45 GHz, with two rectifier circuits, one for each polarization, Figure 2.20(a). The diode impedance at the fundamental is partially pre-matched for lowest insertion loss, and the optimal input impedance required of the patch antenna at 2.45 GHz is found by load-pull to be $16.9 + j5.8$ Ω. The harmonic frequency impedances are shown in Figure 2.20(b), indicating that the 5 harmonics are terminated with low impedances, as desired for efficient operation. The patch antenna feeds are indented 3.6 mm from the center of the patch to achieve this impedance. The two feed points of the antenna are connected to the circuit through metalized vias in the common ground plane. Different substrates are used for the antenna and circuit in order to miniaturize the circuit and keep the antenna efficiency high. Since the rectifier and patch antenna input

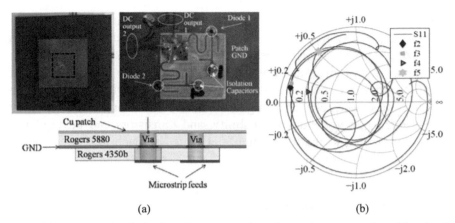

(a)   (b)

Figure 2.20 (a) Dual-polarized patch rectenna, where the patch antenna and rectifier circuit share a ground plane. (b) Measured S_{11} of the rectifier circuit, including diode package inductance, from 2.45 GHz to 12.5 GHz of a single rectifier circuit while port 1 is at the diode plane port 2 is connected to the antenna feed and port 3 (the dc port) is open. The markers are at the second to fifth harmonics of the fundamental, showing close to class-C terminations, which should result in high efficiency (referenced to 50 Ω).

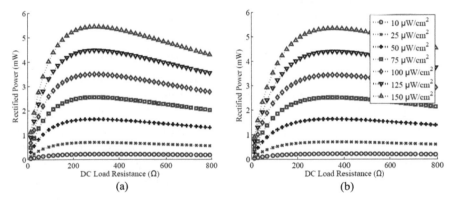

(a)   (b)

Figure 2.21 Measured rectified power versus dc load over a range of incident power densities for vertical (a) and horizontal (b) polarization, both co-polarized with the transmitting antenna. The two rectifier circuits have separate dc loads, predicted to be the best solution for average efficiency in a randomly polarized environment. Each of the measurements was taken with the second dc port unloaded.

impedances match, no additional matching circuit is needed, reducing the insertion loss and size.

Figure 2.21 shows the measured rectified power for the two orthogonal polarizations, showing that the balance is excellent. This will result in the highest average efficiency for randomly-polarized incident waves. The power densities correspond to power levels at the rectifier input that are in the range of 0.1 to a 10 mW (antenna-dependent), from Equation (2.3).

This medium-power dual-polarized rectenna example illustrates the benefit of designing rectifiers in analogy to power amplifier classes of operation. This becomes more relevant at higher power levels, where diodes generally have limited power handling capability, and transistors can be used in a configuration dual to that of an amplifier. This approach to rectifier design is described in the next section.

2.4 High-Power Directive Beaming

To achieve the high power necessary for high-power directive beaming transmitters, two approaches have been considered: solid-state combining in an active phased array and magnetrons. There are many works on solid-state phased array, and specifically spatial combining arrays, e.g. [49]. The beamforming is very relevant for this application [50], and for even higher power levels, magnetron phased-array spatial power combining has been demonstrated [51], as reviewed elsewhere in this text.

2.4.1 Rectifiers for High Power at Microwave Frequencies

At higher power levels (directive beaming), Schottky diodes can break down, and some researchers demonstrated approaches to distribute the power between multiple diodes. In [52], 100-W at 2.45 GHz were rectified with the structure in Figure 2.22(a). A power divider distributes the incident received power to a Schottky diode array, through sub-arrays of 9 rectenna elements, where each rectifier demonstrated efficiency as a function of dc load shown in Figure 2.22(b). A second 16-element rectifier array for 100-W rectification is also shown in [52], where a regulated circuit protects the diodes from breakdown and reduces ripple in the dc output. A rectified power of 67W into a 1.2-Ω load was measured when the array was illuminated by a magnetron at a beaming distance of 5.5 m.

For the beaming approach, higher frequencies are desired to reduce antenna size, and there is a tradeoff between the beaming efficiency (given by the electrical size of the aperture) and the efficiency of rectifier (and PA) circuits at increasing frequencies. In this section, rectifiers at S and X band are discussed to show examples of efficiency decrease for watt-level rectification.

For higher input power rectification, very efficient self-synchronous transistor rectifiers have been demonstrated at microwave frequencies,

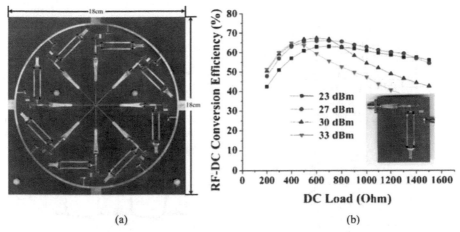

(a)   (b)

Figure 2.22   (a) Schottky diode subarray fed through a power divider for receiving greater than 100 W of input power provided from a magnetron source at 5.5 m distance. (b) The measured conversion efficiency with different dc loads for a single rectifier circuit.

e.g. [48, 53]. It has been shown that under certain loading, a microwave transistor can achieve similar efficiencies as both a power amplifier and a rectifier due to its time-reversal duality property [54]. Assume a PA operating in some high-efficiency mode has an RF output power P_{OUT} at a supply voltage of V_D and appropriate gate drive and bias. If now the drain supply is disconnected and an RF power $P_{IN} = P_{OUT}$ input into the RF drain port, the PA will behave as a rectifier with a conversion efficiency equal to the power-added efficiency of the PA, $\eta = PAE$. Furthermore, the output dc voltage $V_{dc} = V_D$ across some optimal dc load R_D, assuming the gate bias and input drive conditions are kept the same. This is illustrated in Figure 2.23 and is referred to as time-reversal duality, where the drain voltages and currents in the two circuits are the negatives of each other and satisfy: $v_{PA}(t) = v_R(-t)$, and $i_{PA}(t) = -i_R(-t)$.

Synchronous operation of the rectifier requires a second RF source to drive the gate of the transistor to turn it on [55]. Self-synchronous operation relies on power coupled from the drain to the gate through the shared capacitance, C_{gd}. With a highly reflective termination, Z_{gate}, the coupled power can be reflected into the gate to turn on the transistor without a second RF source. Simulations using a unique nonlinear model that includes the third quadrant of the I-V curves [54] results in time-domain waveforms as shown in Figure 2.23(b), but it is shown that this relationship holds true for any class of PA, not just the to-date-demonstrated classes F [54], F^{-1} [48] and E [55–57]. In [48], a theoretical analysis of harmonically terminated

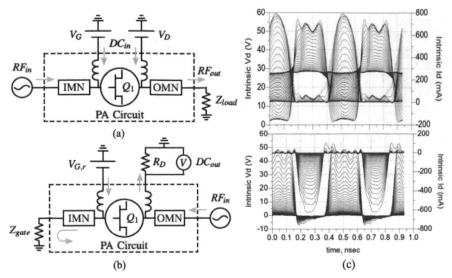

Figure 2.23 PA-rectifier duality: (a) PA circuit and (b) rectifier circuit. In this case, the rectifier is operating self-synchronously with no RF input at the gate. (c) Simulated time-domain waveforms for a class-F PA (*top*) and rectifier (*bottom*), showing time-reversal duality.

high-efficiency power rectifiers based on a Fourier analysis of current and voltage waveforms is presented, in analogy to harmonically terminated PA theory. From the analysis, one can obtain an optimal value for the dc load given the RF circuit design. An upper limit on rectifier efficiency is also derived as a function of device on-resistance.

Some results for demonstrated GaN self-synchronous rectifiers at S and X bands are reviewed here. The 2.14-GHz PA shown in Figure 2.17(a) is designed using the Qorvo TGF2023-02 GaN HEMT with class-F^{-1} harmonic terminations implemented at the second and third harmonics. From the measured data in Figure 2.17(b), at a drain voltage of 28V and a bias current of 160 mA, the PAE = 84% with P_{OUT} = 37.6 dBm and a gain of 15.7 dB under 3-dB compression. The same PA is characterized as a rectifier, with the gate biased and connected to an impedance tuner, converting the two-port transistor PA to a one-port rectifier (Figure 2.23(b)). The measured results are summarized in Figure 2.24, showing the dependence on gate impedance termination Z_{gate} under self-synchronous operation. The rectifier demonstrates an efficiency of 85% for a 10-W input RF power at the transistor drain with a dc voltage of 30V across a 98-Ω resistor. When compared to the PA results, this experimentally validates PA-rectifier duality.

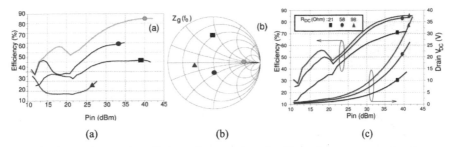

(a)   (b)   (c)

Figure 2.24   (a) Conversion efficiency for several RF load impedances presented at the gate. $V_G = -4.4V$ and $R_{dc} = 98\Omega$. The green point on the Smith chart in (b) corresponds to the highest efficiency point at $Z_{gate} = 230 + j10\Omega$. (c) Conversion efficiency and dc output versus. input power for several dc drain resistor values. $V_G = -4.4V$ and $Z_{gate} = 230 + j10\Omega$. The highest efficiency of 85% is obtained at 40 dBm with a $V_D = 30V$.

To further validate time-reversal duality, two high-efficiency GaN MMIC PAs are characterized under PA and rectifier operation [53] at X-band, Figure 2.25. Both MMICs are designed in Qorvo's 150 nm GaN-on-SiC process. Circuit A is a single-stage amplifier using a 10×100 μm transistor biased at pinch-off ($I_{DQ} = 5$ mA), with an output matching network optimized for efficiency, but with no specific harmonic terminations. Circuit B is a single-stage amplifier that combines two 10×100 μm transistors, biased in deep class-AB, with a reactive combiner. The MMIC performance is summarized in the table as the dc load on the drain is varied along with input power. Similar characterization is performed with gate bias variation and it is found that deep pinch-off improves the efficiency by 12 and 6.2 points in the two rectifiers respectively, but has little effect on the input impedance. The table in Figure 2.25 summarizes the performances of the two X-Band GaN MMICs operating as both amplifiers (PAE) and rectifiers (conversion efficiency). In both modes, the RF power is the power located at the RF drain port, because the duality of these modes states that the output power of the amplifier should be the required input power of the rectifier for peak rectification efficiency. The dc load at the drain is 100 Ω.

The PA-rectifier duality is very general and also applies to two-stage PAs [57]. A 2-stage X-band GaN MMIC PA, biased in class AB, achieves over 10W of output power, >20 dB of saturated gain and a PAE of 50% at 9.9 GHz. Over 52% RF-dc conversion efficiency at a power level of >8W is measured in rectification, Figure 2.26.

There are applications for higher-power efficient microwave rectifiers beyond WPT, for example chip-scale dc–dc converters for power supplies.

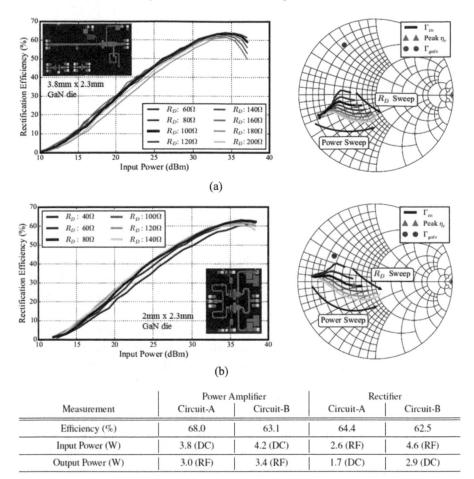

(a)

(b)

Measurement	Power Amplifier		Rectifier	
	Circuit-A	Circuit-B	Circuit-A	Circuit-B
Efficiency (%)	68.0	63.1	64.4	62.5
Input Power (W)	3.8 (DC)	4.2 (DC)	2.6 (RF)	4.6 (RF)
Output Power (W)	3.0 (RF)	3.4 (RF)	1.7 (DC)	2.9 (DC)

Figure 2.25   Measured rectification efficiency and input impedance of the rectifier under input power and dc drain impedance sweeps for (a) single-ended PA (circuit A) and (b) power combined PA (circuit B) at 10.7 GHz. An insignificant ~2.5 point fluctuation occurs in the efficiency with load variation, while the input impedance shows more significant variation in both cases. The table shows a comparison of amplifier and rectifier performance for the two circuits.

Higher switching frequencies are accompanied by reduced efficiency and power since the losses in both passive and active components increase with frequency. Over two decades ago, as high as 64% efficiency was obtained with GaAs devices in a circuit based on transmission lines only, operating at 4.6 GHz at sub-watt power [58], with both a power amplifier and a

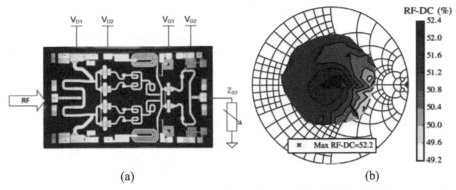

(a)   (b)

Figure 2.26 (a) Two-stage MMIC PA configured as rectifier. (b) Total RF-dc conversion efficiency measured contours showing how the RF load affects system efficiency. A maximum PAE of 52.2% is achieved at Zg = 69-j0.4Ω at 10 GHz.

power oscillator as the inverter stage. In [55] and [56], dc–dc converters implemented with a PA using a 0.25 μm Qorvo GaN die as the dc–AC stage (inverter) and a time-reversed dual rectifier with a resonant dc-isolated coupling network are demonstrated around 1 GHz with 75% efficiency at 5W. The same architecture was integrated in the 150-nm GaN process at 4.6 GHz [59], with a decreased efficiency due to the increased switching losses at this frequency. Nevertheless, this 2.3 mm × 3.8 mm integrated converter is fully monolithic with no magnetic components, Figure 2.27. The total efficiency of around 50% indicates that both rectifier and amplifier are operating at >70% efficiency. In this case, the block diagram in Figure 2.2 still holds, with the middle block reduced to a on-chip resonant circuit.

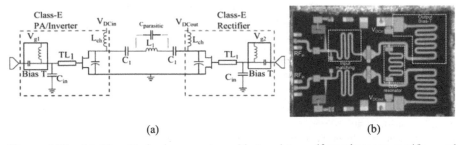

(a)   (b)

Figure 2.27 (a) Class-E dc–dc converter with transistor self-synchronous rectifier and (b) GaN on SiC die operating at 4.6 GHz showing the PA and rectifier in a 2.3 mm × 3.8 mm chip.

2.5 High-Power Near-Field Reactive WPT

High-power near field WPT has many applications, ranging from underwater cable connections to UAV powering and electric vehicle (EV) charging. A possible solution to the limited driving range problem of electric vehicles is wireless powering of the EV in motion on the road, which also reduces battery requirements. Power transfer on the order of 20 kW can be used to power an EV at cruising speeds and is usually the design specification for these systems [60]. Additionally, WPT is an attractive alternative for charging EVs in parking lots, garages and other static scenarios where a wireless system can prove to be more convenient and reliable. So far efforts have been made to demonstrate wireless power transfer to electric vehicles using inductive power transfer, e.g. [60–62], and a few limited cases of capacitive power transfer, as in [63, 64]. Here an example capacitive WPT (CWPT) modular system that can scale to 50 kW over a 12 cm gap using an area of less than $1m^2$ with an efficiency above 85%, illustrated in Figure 2.28. CWPT has several advantages over inductive WPT systems: (1) does not require the use of heavy and expensive ferrites for field concentration; (2) benefits from higher frequencies of operation since the capacitive reactance between the vehicle and the road is inversely proportional to frequency [65]; (3) the displacement current corresponding to high power transfer requires a lower electric field at higher frequencies; and (4) sensitivity to misalignment is reduced using an appropriate geometry.

Due to the large electromagnetic fields that are associated with high power transfer, safety is an important issue that remains to be solved. Exposure

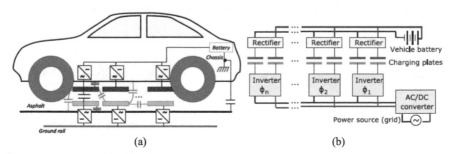

(a)   (b)

Figure 2.28   (a) An overall geometry for a modular CWPT system for electric vehicles. Each module consists of an inverter in the road, and a pair of capacitive plates and a rectifier in the vehicle that charges the on-board battery. Parasitic capacitances from plates to chassis, plate to road-rail, chassis to road-rail and between adjacent plates are also shown. (b) Block diagram of a multi-module CWPT system, with receiving-side capacitor plates shown in orange, and road-side plates shown in green. Phase shifts are applied within the inverters in order to reduce the fringing fields and meet safety requirements.

limits like the ones found in [66] restrict the electromagnetic fields that humans can be exposed to, and these must be met for any product commercialization. Approaches to fringing field reduction using near-field phased arrays of inductive [67, 68] and capacitive [69, 70] powering modules have been previously attempted, and here an example system based [69] is detailed. A photograph of a single module and the corresponding circuit diagram are shown in Figure 2.29. The inverter has an H-bridge topology and is implemented with 650V, 15A GaN on Si FETs (GaNSystem GS66504B). The frequency of operation is 6.78 MHz (ISM band). Power is transfered through a pair of coupling plates, one in the forward path and one in the return path. To minimize circulating current and enable soft-switching in the inverters, it is desired to have a near-resistive input impedance, requiring compensation which is implemented using L-section matching networks at the primary and secondary side.

In order to mimic a practical vehicle charging system, aluminum sheets are added above and below the coupling plates to model the vehicle and the road, as shown in Figure 2.29(a), where the top aluminum sheet has been removed for visibility. The primary and secondary side matching networks are then designed to transfer power through the equivalent series capacitances of this π-network and the parallel capacitances are absorbed into the L-section matching networks. The experimental system has square plates with a side $s = 17.68$ cm, and the L-section matching network has a series capacitance $C_S = 14.2$ pF, a parallel capacitance $C_P = 0.8$ pF and series

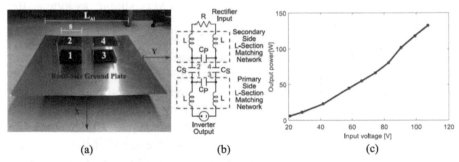

(a)   (b)   (c)

Figure 2.29   (a) Single module photograph showing the four capacitive plates. The top aluminum sheet that models the vehicle has been removed for visibility. The side length of the aluminum sheet is $L_{Al} = 100$ cm. The red arrows indicate the two measurement locations. (b) Equivalent circuit of the CWPT system including the L-section matching networks. CS and CP represent the effective capacitive coupling of the CWPT system. (c) Measured output power of the single module system with versus input dc voltage. Output powers above 110 W correspond with 90% efficiency.

inductors $L = 22.4\,\mu H$, following the design method of [71]. This prototype provides 110 W of output power with an efficiency of 90% to an RF load $R = 42\,\Omega$ that simulates the input of the rectifier. Figure 2.29(c) shows the measured output power as a function of input dc voltage. Efficient inverters and rectifiers for WPT at these power levels have been extensively published in the power electronics community, an example of circuits specifically designed for kW-level wireless power transfer is given in [72]. In wireless power transfer system, typically the efficiency varies with misalignment of transmit and receive coils (for inductive powering) or plates (for capacitive powering). Many publications address this problem for lower power levels, but the impact on the absolute lost power is highest in high-power systems. Approaches that provide some solutions to this problem with adaptive circuits are given in [73] and [74] for inductive and capacitive powering, respectively.

The total magnitude of the electric field vector inside and around the vehicle, where people may be present, has to be below the limits set by the International Commission on Non-Ionizing Radiation Protection (ICNIRP). The electric field produced by the single-module prototype exceeds the ICNIRP safety limit of 33.4 V/m (RMS) at 6.78 MHz and will increase as the output power increases to the required kW levels. To comply with the electric field exposure limits in [66], a multi-module CWPT system can be used (Figure 2.28(b)), where the near field reduction is achieved by a relative phase between adjacent identical modules, analogous to beam steering in the far field of phased array antennas. In [75, 76] it is shown by full-wave simulations that the optimum field cancellation is obtained when the phase shift is $180°$, making the feeding of the modules very simple, as long as the feeding is done in a balanced way.

Full-wave EM simulations of the CWPT system with plate geometry corresponding to the measured module were performed using ANSYS HFSS, Figure 2.30(a) shows the setup used for the simulations. A two-module experimental setup is used with plates 5 cm $\times$ 5 cm in size and with separations of $h = 12$ cm between plates and $d = 15$ cm between pairs of plates of the same module. The modules are located at a distance $D = 20$ cm from each other. A load resistance of 2 kΩ represents the input of the secondary side L-section matching network loaded with the rectifier. The feeding signal is generated using an RF transmitter to emulate the output of the inverter and provide a reasonably high power level. The plates are fed with a 50V peak sine wave generated by a ICOM IC-7410 frequency locked transmitter at 7, 14 and 29 MHz. The voltage applied to the plates is monitored using an oscilloscope with differential probing. The electric field is measured using an ETS HI-6005 electric field probe. The plates are fed using high-power

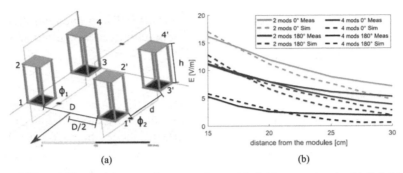

(a)   (b)

Figure 2.30 (a) Simulation setup for comparison with field measurments. (b) E-field measurement (solid) and simulations (dashed) at 7 MHz as a function of distance along x axis for two modules (green and red), four modules (blue and black), and different phases.

baluns (W2AU). Figure 2.30(b) shows the electric field as a function of distance at 7 MHz, while Figure 2.31(a) shows the results at 29 MHz for different relative phases between modules.

Figure 2.31(b) shows the field reduction computed as $\Delta E = (E_0 - E_{180})/E_0$, where E_0 and E_{180} are the magnitudes of the electric field generated by the system when the feeding is done in phase and with phasing of 180°. The field reduction is above 24% for the two-module system and 32% for the four-module system at three frequencies. At closer distances (less than 25 cm), where the fields are more intense, the reduction is above 24% and 43% for the two and four module systems, respectively. The results of the simulations diverge from the measurements at larger distances due to the decreased sensitivity of the field measurement at lower field strengths and

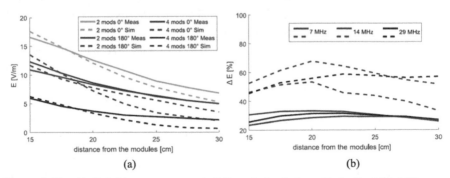

(a)   (b)

Figure 2.31 (a) E-field measurement (solid) and simulations (dashed) at 29 MHz as a function of distance along x axis for two modules (green and red), four modules (blue and black), and different phases. (b) Two and four module systems electric field difference (ΔE) between the 0° and the 180° phase shifted feed; the solid and dashed traces correspond to the two and four module systems, respectively.

lack of modeling of surrounding objects. This CWPT approach has recently been scaled by the same group to 1 kW system with over 85% efficiency and over 60% reduction in emitted fields.

2.6 Conclusion

This chapter presents some approaches to solid-state circuits for wireless power transfer. It covers examples from MHz-frequency high-power capacitive transfer for electric vehicles to GHz frequency WPT in the µW and mW power range. Transmitters are discussed briefly, and the emphasis is placed on power receivers, since these are specific to wireless powering applications.

References

[1] Z. Popovic, "Cut the Cord: Low-Power Far-Field Wireless Powering," in *IEEE Microwave Magazine*, vol. 14, no. 2, pp. 55–62, March–April 2013.

[2] A. Collado and A. Georgiadis, et al., "Improving wireless power transmission efficiency using chaotic waveforms," *2012 IEEE MTT-S International Microwave Symposium Digest*, Montreal, Canada, June 2012.

[3] Z. Popovic, E. Falkenstein, D. Costinett and R. Zane, "Low-power far-field wireless powering for wireless sensors," *Proceedings of the IEEE, Special Issue on Wireless Powering*, vol. 101, no. 6, pp. 1397–1409, June 2013.

[4] R. Vias, H. Nishimoto, M. Tentzeris, Y. Kawahara and T. Asami, "A Battery-Less, Energy Harvesting Device for Long Range Scavenging of Wireless Power from Terrestrial TV Broadcasts," *IEEE 2012 IMS Digest*, Montreal, Canada, June 2012.

[5] R. Scheeler, S. Korhummel and Z. Popovic, "A Dual-Frequency Ultralow-Power Efficient 0.5-g Rectenna," in *IEEE Microwave Magazine*, vol. 15, no. 1, pp. 109–114, Jan.–Feb. 2014.

[6] T. Paing, J. Shin, R. Zane and Z. Popovic, "Resistor emulation approach to low-power RF energy harvesting," *IEEE Trans. Power Electronics*, vol. 23, no. 3, May 2008, pp. 1494–1501.

[7] E. Falkenstein, M. Roberg and Z. Popović, "Low-power wireless power delivery," *IEEE Trans. Microwave Theory Techn.*, vol. 60, no. 7, pp. 2277–2286, July 2012.

[8] A. Dolgov, R. Zane and Z. Popovic, "Power management system for online low power RF energy harvesting optimization," *IEEE Trans. Circuits Syst.*, vol. 57, no. 7, pp. 1802–1811, Jul. 2010.

[9] M. D. Seeman, S. R. Sanders and J. M. Rabaey, "An ultra-low-power power management IC for wireless sensor nodes," in *Proc. IEEE 29th Custom Integr. Circuits Conf.*, San Jose, CA, USA, Sep. 2007, pp. 567–570.

[10] I. Doms, P. Merken, R. P. Mertens and C. Van Hoof, "Capactive power-management circuit for micropower thermoelectric generators with a 2.1 μW controller," in *Proc. IEEE Int. Solid-State Circ. Conf.*, San Francisco, CA, USA, Feb. 2008, pp. 300–303.

[11] E. E. Aktakka, R. L. Peterson and K. Najafi, "A self-supplied inertial piezoelectric energy harvester with power-management IC," *Intern. Solid-State Circuits Conference (ISSCC)*, Feb. 2011, pp. 120–121.

[12] T. Paing, J. Shin, R. Zane and Z. Popovic, "Custom IC for Ultralow Power RF Energy Scavenging," *IEEE Trans. on Power Electron., Letters*, vol. 26, no. 6, pp. 1620–1626, Jun. 2011.

[13] E. Falkenstein, D. Costinett, R. Zane and Z. Popovic, "Far-field RF-powered variable duty cycle wireless sensor platform," *IEEE Trans, Circuits and Systems II*, vol. 58, no. 12, pp. 822–826, Dec. 2011.

[14] Ramos and Z. Popovic, "A compact 2.45 GHz, low power wireless energy harvester with a reflector-backed folded dipole rectenna," in *Wireless Power Transfer Conference (WPTC)*, 2015 IEEE, pp. 1–3, May 2015.

[15] G. Orecchini, L. Yang, M. M. Tentzeris and L. Roselli, "Wearable Battery-free Active Paper-Printed RFID Tag with Human Energy Scavenger," *IEEE Intern. Microwave Symp. Digest*, Baltimore, MD, June 2011.

[16] Z. Popovic, S. Korhummel, S. Dunbar, R. Scheeler, A. Dolgov, R. Zane, E. Falkenstein and J. Hagerty, "Scalable rf energy harvesting," *IEEE Trans. Microwave Theory and Techn.*, vol. 62, pp. 1046–1056, April 2014.

[17] J. A. Hagerty, F. Helmbrecht, W. McCalpin, R. Zane and Z. Popovic, "Recycling ambient microwave energy with broadband antenna arrays," *IEEE Trans. Microwave Theory and Techn.*, pp. 1014–1024, March 2004.

[18] U. Olgun, C. C. Chen and J. L. Volakis, "Investigation of Rectenna Array Configurations for Enhanced RF Power Harvesting," in *IEEE Antennas and Wireless Propagation Letters*, vol. 10, pp. 262–265, 2011.

[19] A. Boaventura, D. Belo, R. Fernandes, A. Collado, A. Georgiadis and N. B. Carvalho. "Boosting the efficiency: Unconventional waveform design for efficient wireless power transfer," in *IEEE Microwave Magazine*, vol. 16, no. 3, pp. 87–96, 2015.

[20] J. Estrada, I. Ramos, A. Narayan, A. Keith and Z. Popovic, "RF energy harvester in the proximity of an aircraft radar altimeter," *2016 IEEE Wireless Power Transfer Conference (WPTC)*, Aveiro, 2016, pp. 1–4.

[21] C. R. Valenta and G. D. Durgin, "Harvesting Wireless Power: Survey of Energy-Harvester Conversion Efficiency in Far-Field, Wireless Power Transfer Systems," in *IEEE Microwave Magazine*, vol. 15, no. 4, pp. 108–120, June 2014.

[22] J. Kimionis, A. Collado, M. M. Tentzeris and A. Georgiadis, "Octave and Decade Printed UWB Rectifiers Based on Nonuniform Transmission Lines for Energy Harvesting," in *IEEE Transactions on Microwave Theory and Techniques*, vol. PP, no. 99, pp. 1–9.

[23] C. Song, Y. Huang, J. Zhou, J. Zhang, S. Yuan and P. Carter, "A High-Efficiency Broadband Rectenna for Ambient Wireless Energy Harvesting," in *IEEE Transactions on Antennas and Propagation*, vol. 63, no. 8, pp. 3486–3495, Aug. 2015.

[24] S. E. Adami, P. Proynov, G. S. Hilton, G. Yang, C. Zhang, D. Zhu, Y. Li, S. P. Beeby, I. J. Craddock and B. H. Stark, "A Flexible 2.45-GHz Power Harvesting Wristband With Net System Output From -24.3 dBm of RF Power," in *IEEE Transactions on Microwave Theory and Techniques*, vol. PP, no. 99, pp. 1–16.

[25] S. Kim et al., "Ambient RF Energy-Harvesting Technologies for Self-Sustainable Standalone Wireless Sensor Platforms," in *Proceedings of the IEEE*, vol. 102, no. 11, pp. 1649–1666, Nov. 2014.

[26] C. H. P. Lorenz et al., "Breaking the Efficiency Barrier for Ambient Microwave Power Harvesting With Heterojunction Backward Tunnel Diodes," in *IEEE Transactions on Microwave Theory and Techniques*, vol. 63, no. 12, pp. 4544–4555, Dec. 2015.

[27] C. H. P. Lorenz, S. Hemour and K. Wu, "Physical Mechanism and Theoretical Foundation of Ambient RF Power Harvesting Using Zero-Bias Diodes," in *IEEE Transactions on Microwave Theory and Techniques*, vol. 64, no. 7, pp. 2146–2158, July 2016.

[28] http://powermat.com, http://www.qualcomm.com/solutions/wireless-charging

[29] S. Aldhaher, D. C. Yates and P. D. Mitcheson, "Design and Development of a Class EF2 Inverter and Rectifier for Multimegahertz Wireless Power Transfer Systems," in *IEEE Transactions on Power Electronics*, vol. 31, no. 12, pp. 8138–8150, Dec. 2016.

[30] D. Ahn and P. P. Mercier, "Wireless power transfer with concurrent 200 kHz and 6.78 MHz operation in a single transmitter device," *IEEE Trans. Power Electron.*, vol. 31, no. 7, pp. 5081–5029, Jul. 2016.

[31] S. Aldhaher, P. C.-K. Luk, K. El Khamlichi Drissi and J. F. Whidborne, "High-input-voltage high-frequency Class E rectifiers for resonant inductive links," *IEEE Trans. Power Electron.*, vol. 30, no. 3, pp. 1328–1335, Mar. 2015.

[32] Hui, S. Y. Ron and Wing W. C. Ho, "A new generation of universal contactless battery charging platform for portable consumer electronic equipment," *IEEE Transactions on Power Electronics* 20.3 (2005): 620–627.

[33] S. Korhummel, A. Rosen and Z. Popović, "Over-Moded Cavity for Multiple-Electronic-Device Wireless Charging," *IEEE Transactions on Microwave Theory and Techniques*, vol. 62, no. 4, pp. 1074–1079, April 2014.

[34] A. G. Zajić and Z. Popović, "Statistical modeling of a shielded wireless charging device," *2015 9th European Conference on Antennas and Propagation (EuCAP)*, Lisbon, 2015, pp. 1–5.

[35] F. Raab, P. Asbeck, S. Cripps, P. Kenington, Z. Popovic, N. Pothecary, J. Sevic and N. Sokal, "Power amplifiers and transmitters for RF and microwave," *IEEE transactions on Microwave Theory and Techniques* 50, no. 3 (2002): 814–826.

[36] M. Roberg, J. Hoversten and Z. Popović, "GaN HEMT PA with over 84% power added efficiency," *Electron. Lett.*, vol. 46, no. 23, pp. 1553–1554, Nov. 2010.

[37] S. Schafer, M. Litchfield, A. Zai, Z. Popović and C. Campbell, "X-Band MMIC GaN Power Amplifiers Designed for High-Efficiency Supply-Modulated Transmitters," *IEEE MTT International Microwave Symp. Digest*, June 2013, Seattle.

[38] S. Piotrowicz, Z. Ouarch, E. Chartier, R. Aubry, G. Callet, D. Floriot, J. Jacquet, O. Jardel, E. Morvan, T. Reveyrand, N. Sarazin and S. Delage, "43W, 52% PAE X-Band AlGaN/GaN HEMTs MMIC Amplifiers," *IEEE MTT-S International Microwave Symposium Digest (IMS)*, 2010, pp. 1–4.

[39] D. Costinett, E. Falkenstein, R. Zane and Z. Popovic, "RF-powered variable duty cycle wireless sensor," *Microwave Conference (EuMC), 2010 European*, pp. 41–44, 2010.

[40] T. Paing, A. Dolgov, J. Shin, J. Morroni, J. Brannan, R. Zane and Z. Popovic, "Wirelessly powered wireless sensor platform," *European Microwave Conf. Digest*, pp. 241–244, Munich, Oct. 2007.

[41] H.-K. Chiou and I.-S. Chen, "High-efficiency dual-band on-chip rectenna for 35- and 94-GHz wireless power transmission in 0.13-um

CMOS technology," *IEEE Trans. Microwave Theory Techn.*, vol. 58, no. 12, pp. 3598–3606, Dec. 2010.

[42] J. J. Schlesak, A. Alden and T. Ohno, "A microwave powered high altitude platform," *Microwave Symposium Digest, 1988., IEEE MTT-S International*, pp. 283–286 vol. 1, 25–27 May 1988.

[43] A. Georgiadis, G. Andia and A. Collado, "Rectenna design and optimization using reciprocity theory and harmonic balance analysis for electromagnetic (em) energy harvesting," *Antennas and Wireless Prop. Letters, IEEE*, vol. 9, pp. 444–446, 2010.

[44] Z. Harouni, L. Cirio, L. Osman, A. Gharsallah and O. Picon, "A dual circularly polarized 2.45-GHz rectenna for wireless power transmission," *Antennas and Wireless Prop. Lett., IEEE*, vol. 10, pp. 306–309, 2011.

[45] D.-G. Youn, K.-H. Kim, Y.-C. Rhee, S.-T. Kim and C.-C. Shin, "Experimental development of 2.45 GHz rectenna using FSS and dual-polarization," *Microwave Conf. 30th European*, pp. 1–4, Oct. 2000.

[46] S. Imai et al., "Efficiency and harmonics generation in microwave to DC conversion circuits of half-wave and full-wave rectifier types," in *2011 IEEE MTT-S International*, May 2011, pp. 15–18.

[47] H. Takhedmit et al., "A 2.45-GHz low cost and efficient rectenna," in *Antennas and Propagation (EuCAP), 2010 Proceedings of the Fourth European Conference on*, April 2010, pp. 1–5.

[48] M. Roberg, T. Reveyrand, I. Ramos, E. A. Falkenstein and Z. Popovic, "High-Efficiency Harmonically Terminated Diode and Transistor Rectifiers," in *IEEE Transactions on Microwave Theory and Techniques*, vol. 60, no. 12, pp. 4043–4052, Dec. 2012.

[49] R. A. York and Z. Popovic, eds. *Active and quasi-optical arrays for solid-state power combining.* vol. 42. Wiley-Interscience, 1997.

[50] N. Shinohara and H. Matsumoto, "Experimental study of large rectenna array for microwave energy transmission," *IEEE Transaction MTT*, vol. 46, no. 3, pp. 261–267, Mar. 1998.

[51] N. Shinohara, H. Matsumoto and K. Hashimoto, "Solar Power Station/ Satellite (SPS) with Phase Controlled Magnetrons," *IEICE Trans. Electron,* vol. E86-C, no. 8, pp. 1550–1555, 2003.

[52] Wan Jiang, B. Zhang, Liping Yan and C. Liu, "A 2.45 GHz rectenna in a near-field wireless power transmission system on hundred-watt level," *2014 IEEE MTT-S International Microwave Symposium (IMS2014)*, Tampa, FL, 2014, pp. 1–4.

[53] M. Litchfield, T. Reveyrand and Z. Popovic, "High-efficiency X-band MMIC GaN power amplifiers operating as rectifiers," *2014 IEEE IMS*, June 2014, pp. 1–4.

[54] T. Reveyrand, I. Ramos and Z. Popovic, "Time-reversal duality of high efficiency RF power amplifiers," *Electronics Lett.*, vol. 48, pp. 1607–1608, Dec. 2012.

[55] J. A. Garcia, et al. "GaN HEMT Class E^2 Resonant Topologies for UHF DC/DC Power Conversion," *IEEE Trans. Microwave Theory Techn.*, vol. 60, pp. 4220–4229, Dec. 2012.

[56] I. Ramos et al., "GaN Microwave DC-DC Converters," *IEEE Trans. Microwave Theory Techn.*, vol. 63, pp. 4473–4482, Dec. 2015.

[57] M. Coffey, S. Schafer and Z. Popović, "Two-stage high-efficiency X-Band GaN MMIC PA/rectifier," *2015 IEEE MTT-S International Microwave Symposium*, Phoenix, AZ, 2015, pp. 1–4.

[58] S. Djukic et al. "A planar 4.5-GHz DC–DC power converter," *IEEE Trans. Microw. Theory Techn.*, vol. 47, no. 8, pp. 1457–1460, Aug. 1999.

[59] I. Ramos et al., "A Microwave Monolithically Integrated Distributed 4.6 GHz DC-DC Converter," *IEEE IMS* 2016, San Francisco, June 2016.

[60] G. A. Covic and J. T. Boys, "Modern trends in inductive power transfer for transportation applications," *IEEE Journal of Emerging and Selected Topics in Power Electronics*, vol. 1, no. 1, pp. 28–41, March 2013.

[61] S. Y. R. Hui, W. Zhong and C. K. Lee, "A critical review of recent progress in mid-range wireless power transfer," *IEEE Transactions on Power Electronics*, vol. 29, no. 9, pp. 4500–4511, Sept. 2014.

[62] J. Enriquez, "Qualcomm wireless technology charges electric vehicles in motion," https://www.rfglobalnet.com/, accessed: 2017-05-19.

[63] H. Zhang, F. Lu, H. Hofmann, W. Liu and C. C. Mi, "A four-plate compact capacitive coupler design and lcl-compensated topology for capacitive power transfer in electric vehicle charging application," *IEEE Transactions on Power Electronics*, vol. 31, no. 12, pp. 8541–8551, Dec. 2016.

[64] J. Dai and D. C. Ludois, "Capacitive power transfer through a conformal bumper for electric vehicle charging," *IEEE Journal of Emerging and Selected Topics in Power Electronics*, vol. 4, no. 3, pp. 1015–1025, Sept. 2016.

[65] A. Sepahvand, A. Kumar, K. K. Afridi and D. Maksimovic, "High Power Transfer Density and High Efficiency 100 MHz Capacitive Wireless Power Transfer System," *Proc. IEEE Workshop on Control and Modeling for Power Electronics (COMPEL)*, Vancouver, Canada, July 2015.

[66] I. C. on Non-Ionizing Radiation Protection (ICNIRP), "Guidelines for limiting exposure to time-varying electric, magnetic and electromagnetic fields (up to 300 GHz)," *Health Physics*, vol. 74, no. 4, pp. 494–522, 1998.

[67] B. H. Waters, B. J. Mahoney, V. Ranganathan and J. R. Smith, "Power delivery and leakage field control using an adaptive phased array wireless power system," *IEEE Transactions on Power Electronics*, vol. 30, no. 11, pp. 6298–6309, Nov. 2015.

[68] G. Sauerlaender and E. Waffenschmidt, "Wireless power transmission system," Aug. 19 2014, uS Patent 8,810,071. [Online]. Available: https://www.google.ch/patents/US8810071

[69] A. Kumar, S. Pervaiz, Chieh-Kai Chang, S. Korhummel, Z. Popovic and K. K. Afridi, "Investigation of power transfer density enhancement in large air-gap capacitive wireless power transfer systems," in *Wireless Power Transfer Conference (WPTC), 2015 IEEE*, vol., no., pp. 1–4, 13–15 May 2015, Boulder, CO, U.S.A.

[70] F. Lu, H. Zhang, H. Hofmann and C. Mi, "A Double-Sided LCLC Compensated Capacitive Power Transfer System for Electric Vehicle Charging," *IEEE Transactions on Power Electronics*, vol. 30, no. 11, pp. 6011–6014, November 2015.

[71] S. Sinha, A. Kumar, S. Pervaiz, B. Regensburger and K. K. Afridi, "Design of efficient matching networks for capacitive wireless power transfer systems," in *2016 IEEE 17th Workshop on Control and Modeling for Power Electronics (COMPEL)*, June 2016, pp. 1–7, Trondheim, Norway.

[72] J. Choi, D. Tsukiyama, Y. Tsuruda and J. Rivas, "13.56 MHz 1.3 kW resonant converter with GaN FET for wireless power transfer," *Proc. IEEE Wireless Power Transfer Conf.*, May 2015, pp. 1–4.

[73] C. Florian, F. Mastri, R. P. Paganelli, D. Masotti and A. Costanzo, "Theoretical and Numerical Design of a Wireless Power Transmission Link With GaN-Based Transmitter and Adaptive Receiver," *IEEE Transactions on Microwave Theory and Techniques*, vol. 62, no. 4, pp. 931–946, April 2014.

[74] S. Sinha, A. Kumar and K. K. Afridi, "Active Variable Reactance Rectifier – A New Approach to Compensating for Coupling Variations in Wireless Power Transfer Systems," *Proceedings of the IEEE Workshop on Control and Modeling for Power Electronics (COMPEL)*, Stanford, CA, July 2017.

[75] I. Ramos, K. Afridi, J. A. Estrada and Z. Popovic, "Near-field capacitive wireless power transfer array with external field cancellation," in *2016 IEEE Wireless Power Transfer Conference (WPTC)*, May 2016, pp. 1–4, Aveiro, Portugal.

3

Microwave Tube Transmitters

Tomohiko Mitani

Kyoto University, Japan

Abstract

Microwave tube transmitters (magnetrons, klystrons and amplitrons) are described in this chapter. In particular, the magnetrons are still used as the transmitters of microwave power transfer studies. Injection locking magnetrons, phase-controlled magnetrons, phase-and-amplitude-controlled magnetrons, and power-variable phase controlled magnetrons are described in detail as highly-functionalized magnetron transmitters. Demonstration experiments including future concepts on microwave power transfer by using the microwave tube transmitters are also introduced.

3.1 Introduction

Microwave tubes are vacuum tubes which generate or amplify radio-frequency (RF) electromagnetic waves in the microwave band. The history of the microwave tubes dates back to the early 20th century. Barkhausen and Kurz observed self-excited vibration called "electron dance oscillation" in a triode tube and succeeded in extracting high-frequency electromagnetic waves at the shortest wavelength of 43 cm [1]. Almost in the same age, a magnetron was invented by Hull [2] and improved in oscillation efficiency in the microwave band by splitting the anode by Okabe [3]. In the middle of 20th century, klystrons were independently invented by two research groups [4, 5]. Also a traveling wave tube (TWT) was first invented by Haeff [6] and studied theoretically and experimentally [7–9]. Other types of microwave tubes such as cross-field amplifiers (CFAs), gyrotrons, etc. were subsequently invented in the latter part of the 20th century [10, 11].

49

Features of the microwave tubes are high output power, high voltage operation and thermal tolerance compared with solid-state devices. Thanks to high output power, the microwave tubes are still used for radar applications, microwave heating applications, transmitters of satellite communications, etc. In particular, magnetrons become an essential element of home appliances, *i.e.* a microwave oven.

In this chapter, microwave tube transmitters for wireless power transfer (WPT), especially microwave power transfer (MPT), are introduced. The excellent progress of MPT in the 20th century is largely due to the diverse and sophisticated works of Brown, Raytheon. Since the last quarter of the 20th century, contributions from other countries, especially Japan, have increased.

3.2 Magnetron

A magnetron is a cross-field device and an oscillator mainly used for microwave heating and radar applications. It is basically designed in the ISM (Industrial, Scientific and Medical) bands for microwave heating. In particular, a 2.45 GHz oven magnetron is still the cheapest 1–10 kW class microwave oscillator in the world despite being invented almost a century ago. A 915 MHz magnetron is often used for industrial microwave heating applications, though the frequency allocation of the 915 MHz ISM band is confined to north and south American countries. Regarding the 5.8 GHz ISM band, an over-1 kW magnetron was recently produced [12].

Thanks to their availability, efficiency and cost, 2.45 GHz oven magnetrons have been often used as the transmitters of MPT studies. Various types of magnetron transmitters with frequency locking, phase locking/control and output power control have been developed. 5.8 GHz CW magnetrons have also been used for some MPT studies.

3.2.1 Operating Principles

A magnetron has a coaxial diode structure as its cross-sectional view is shown in Figure 3.1. The cathode in the center plays a role of filament which emits thermal electrons. The anode has multiple cavity resonators, the number of which is even in most magnetrons. As the electric field E is applied between the anode and the cathode, the external magnetic field B is applied in the axial direction as shown in Figure 3.1. Then the electrons move in the azimuth direction by $E \times B$ drift.

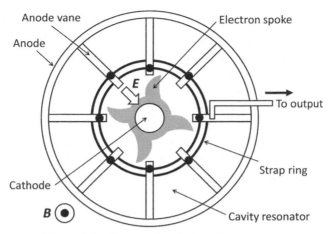

Figure 3.1   Cross-sectional view of a magnetron.

Since the anode cavities exhibit various resonant modes, small RF fluctuations exist in the interaction region between the anode and the cathode. When the azimuthal electron motion is synchronized with one of the resonant mode, the electrons gather around the anode vanes with the positive RF potential, and escape from the anode vanes with the negative RF potential, due to the $E \times B$ drift effect. Finally, the electrons build "spokes" in the interaction region with the azimuth drift motion, while the RF field is strongly enhanced and oscillated at the resonant frequency. The oscillating RF power is extracted from the output antenna attached to one of the anode vanes as shown in Figure 3.1, or the output window opened at the anode.

The magnetron operation condition is restricted by the following Buneman-Hartree threshold voltage V_{th},

$$V_{th} = \frac{\omega_m B}{2n} \left(r_a^2 - r_c^2 \right) - \frac{m \omega_m^2 r_a^2}{2en^2} \tag{3.1}$$

where B, r_a, r_c, ω_m, e and m are the external magnetic flux density, the anode radius, the cathode radius, the oscillation angular frequency, the elementary charge and the electron mass, respectively. n is an integer from 1 to $N/2$, where N is the number of anode vanes (N is usually the even number). When the anode voltage V_a is larger than V_{th}, the magnetron oscillation starts. When $n = N/2$, the phase difference of the RF potential between the adjacent anode vanes becomes π. This oscillation condition is called "π mode". As V_{th} becomes the minimum for the π mode from Equation (3.1), most magnetrons are designed under the π mode condition. To strengthen the π mode oscillation, the strap rings are usually attached to the anode vanes as shown in Figure 3.1. Since a strap ring is electrically connected to every other vane, the phase difference between the adjacent vanes prefers π.

The RF-dc conversion efficiency η of a magnetron was expressed as $\eta = \eta_c \eta_e$, where η_c is a circuit efficiency and η_e is an electronic efficiency, respectively. η_c is determined by the quality factors in the magnetron output circuit, that is, $\eta_c = Q_L/Q_E$ where Q_L is the load Q and Q_E is the external Q, respectively [13]. η_e is generally determined by the input energy (potential energy) and the residual kinetic energy of electrons. It is expressed differently depending on the assumption of electron density in the magnetron. The following η_e is one of the analytical expressions [11].

$$\eta_e = 1 - \frac{1}{2V_a}\left\{ \frac{B(r_a^2 - r_c^2)\omega_m}{n} - \frac{m}{e}\left(\frac{r_a\omega_m}{n}\right)^2\right\} \tag{3.2}$$

$$= 1 - \frac{4m\omega_m}{eBn}\frac{r_a^2}{r_a^2 - r_c^2} + \left(\frac{2m\omega_m}{eBn}\frac{r_a^2}{r_a^2 - r_c^2}\right)^2 \tag{3.3}$$

$$V_a = \frac{eB^2}{8m}\left(\frac{r_a^2 - r_c^2}{r_a}\right)^2 \tag{3.4}$$

η_e becomes the maximum for the π mode from Equation (3.2).

3.2.2 Noise Reduction Methods for an Oven Magnetron

It is often said a magnetron is a noisy oscillator. One of the major reasons is because the anode cavity resonators provide resonance frequencies in various frequency bands as well as the oscillation frequency. Another major reason is attributed to the operation way of the magnetron. In a microwave oven, the magnetron is usually driven by a cheap power supply such as a half-wave voltage doubler or a full-bridge rectifier without a smoothing circuit. Then the magnetron oscillation frequency is varied widely because commonly-used oven magnetrons have a "frequency pushing" effect [10]; the oscillation frequency monotonically increases with the anode current. Figure 3.2 (a) shows a typical frequency spectrum of an oven magnetron driven by a half-wave voltage doubler. With such a wide spectrum, a magnetron is quite difficult to be used for MPT applications.

In 1980s, Brown discovered a method to operate a magnetron quietly, though he used to be skeptical of using an oven magnetron for MPT due to

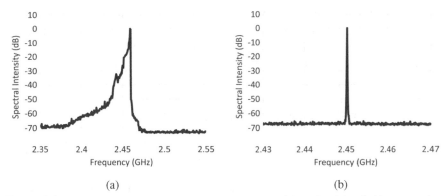

Figure 3.2 Frequency spectra of an oven magnetron driven by (a) a half-wave voltage doubler power supply, and (b) a dc power supply after turning off the filament power.

its reputation of noisiness [14]. When it is driven by a dc power supply with turning off the filament power, the oven magnetron exhibits an extremely low spurious noise level due to the "internal feedback mechanism" [15]. Under a normal operation of the oven magnetron, the cathode filament is continuously heated by an external power supply for thermionic electron emission. However the electron emission still maintains even after turning off the filament power, owing to backbombardment energy. Then the cathode filament temperature autonomously decreases to flow the needed anode current, and finally the magnetron becomes quiet. The noise reduction effect by turning off the filament power is attributed to not only the internal feedback mechanism but also elimination of the anode current fluctuation [16]. The anode current of an oven magnetron becomes unstable in a certain current range where the anode voltage becomes almost constant. After turning off the filament power, the voltage-current characteristics have a slope then the dc power supply can control the anode current stably. Figure 3.2 (b) shows a typical frequency spectrum of an oven magnetron driven by a dc power supply after turning off the filament power. This sharp spectrum is much more suitable for MPT applications. In addition, the magnetron spurious noise outside the oscillation frequency bands can be also reduced by turning off the filament current [17].

3.2.3 Ingection Locked Magnetron

An injection locked magnetron is a magnetron whose oscillation frequency is locked with that of the reference signal by an injection locking method.

Frequency locking is a well-known phenomenon observed in oscillators [18]. In the injection locked magnetron case, the reference signal is injected from the magnetron output antenna via a circulator. When the reference signal frequency is set close to the free-running frequency of magnetron, the output frequency is synthesized with the reference signal frequency. When the magnetron frequency is locked with the reference, the locking range Δf is expressed as the following Adler's equation [10].

$$\frac{\Delta f}{f} = \frac{1}{Q_E} \sqrt{\frac{P_i}{P_o}} \qquad (3.5)$$

where f, Q_E, P_i and P_o are the center frequency of the locking range, the external Q of the magnetron, the injected power, and the magnetron output power, respectively. The injection locked magnetron is often called "magnetron directional amplifier" [15], since it looks like an amplifier to the reference signal while the frequency locking is maintained.

The injection locking method is effective to unify the output frequency of magnetrons whose free-running frequencies are individually different. However, the phase difference between the magnetron output and the reference signal still remains even after the injection locking, because the original free-running frequency of the magnetron is not changed by the external signal injection. The residual phase difference θ is expressed as the following equation,

$$\sin \theta = \frac{\Delta f' Q_E}{f} \sqrt{\frac{P_o}{P_i}} \qquad (3.6)$$

where $\Delta f'$ is the difference between the reference signal frequency and the free-running frequency of the magnetron [10]. Hence, θ will be necessary to be converged to a certain value (*e.g.* zero) when a phased array is constructed by the injection locked magnetrons.

3.2.4 Phase-Controlled Magnetron

A phase-controlled magnetron or phase-locked magnetron (PCM) is a magnetron whose frequency and output phase are both locked to those of the reference signal. A typical schematic diagram of the PCM is shown in Figure 3.3. The frequency and phase locking of PCM are realized by the combination of an injection locking method and a phase-locked loop (PLL) method. As with the injection locked magnetron, the reference signal is injected to the magnetron via a circulator. As described in Section 3.2.3, the

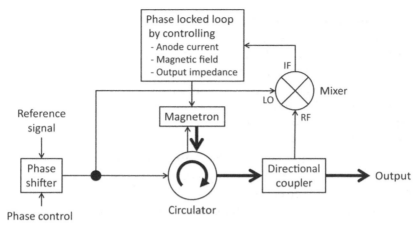

Figure 3.3 Schematic diagram of a phase-controlled magnetron (PCM).

injection locking method only works effectively under the condition that the magnetron frequency is close to the reference frequency. In addition, the magnetron output phase is not determined uniquely according to the difference between the reference signal frequency and the free-running frequency of the magnetron, as expressed in Equation (3.6). The PLL method is therefore adopted to control the magnetron frequency and phase.

In the PLL, the magnetron is regarded as a voltage-controlled oscillator (VCO). The magnetron output phase is compared with the reference signal phase by a mixer. A double-balanced mixer is used for direct phase comparison; whereas a digital phase frequency detector is also used by adopting frequency dividers to the magnetron output and the reference [19, 20]. The IF port of the mixer outputs a sinusoidal wave corresponding to the frequency difference between the magnetron and the reference signal when the magnetron is not locked to the reference. Once the magnetron is locked, the IF port outputs a dc voltage corresponding to the phase difference between the magnetron and the reference signal. Then the magnetron output phase can be locked to the reference signal phase by controlling the magnetron frequency with a feedback loop circuit.

The controlling methods of the magnetron frequency and phase in the PCM are classified into three types: anode current control [19–21], external magnetic field control [15], and output load control [22, 23]. The anode current control utilizes a frequency pushing effect of the magnetron. Since the magnetron frequency is monotonously varied by the anode current in a certain current range, the output phase difference from the reference can be converged to zero by being fed back to the anode current of the magnetron power supply. The external magnetic field control utilizes the frequency variation by controlling the azimuthal drift velocity of electron spokes. Electromagnets are used to control the magnetic field applied to the magnetron. As the phase variation is monotonous against the magnetic field, the output phase difference from the reference can be converged to zero by being fed back to the power supply of the electromagnets. The output load control utilizes a frequency pulling effect of the magnetron. A variable reactance such as a stub tuner can apply to change the output load impedance, which affect the magnetron frequency. By being fed back to the variable reactance operation, the output phase difference from the reference can be set to zero.

The phase shifter is used for controlling the PCM output phase after the frequency and phase locking. By using multiple PCM transmitters, one can construct a high-power phased array antenna. Here the phase convergence of each PCM is extremely important to create a precise microwave beam by the PCM phased array.

Note that these three control methods affect not only the magnetron frequency but also the output power. When the magnetron frequency and phase are locked to the reference, the output power is also fixed. Hence, advanced types of the PCM were developed to have an additional function of output power control, as described in the following sections.

As an optional PCM application, it is reported that the anode current control type PCM can be used as a frequency-modulated and phase-modulated transmitter [20]. The data transmission at 2 Mbps by phase-shift-keying (PSK) has been achieved by the PCM. The data transmission at 250 kbps by frequency-shift-keying (FSK) has been also achieved. Also a pulse-driven 2.45GHz PCM was developed [24]. During the 1 kHz-cycle pulse operation with a duty ratio of 0.5, the output phase of the developed PCM was locked within 100 μs under the injected reference signal power of 4 W. The phase stability of the pulse-driven PCM was less than $\pm 2.4°$.

Regarding a 5.8 GHz magnetron, a compact microwave energy transmitter called COMET was developed in Japan [25]. The COMET consists of an anode current control type PCM at 5.8 GHz, a circular-polarized radial-line-slot antenna as the transmitting antenna, and a heat radiation system. Its size and microwave output were 310 mm-diameter, 99 mm-depth, and 280 W, respectively. Its weight was below 7 kg.

3.2.5 Phase-and-Amplitude-Controlled Magnetron

A phase-and-amplitude-controlled magnetron (PACM) is a magnetron whose frequency and output phase are both locked to those of the reference signal, and whose output power is locked to the reference power. A schematic diagram of the PACM is shown in Figure 3.4. The frequency and phase locking method of the PACM are the same as the PCM with the anode current control PLL. Besides, the external magnetic field control is used for the output power control. A part of magnetron output power is extracted from the directional coupler and rectified to dc. The dc voltage is compared with the reference voltage. Then the output power can be converged to the setting power by being fed back to the power supply of the electromagnets.

A 2.45 GHz PACM was demonstrated in Japan [25]. The developed PACM could change the microwave output from 300 W to 500 W with an output phase stability of less than 1°. Its frequency stability achieved less than 10^{-6}.

3.2.6 Power-Variable Phase-Controlled Magnetron

A power-variable phase-controlled magnetron (PVPCM) is a magnetron whose frequency and output phase are both locked to those of the reference signal, and whose output power can be changed freely without a feedback loop. A schematic diagram of the PVPCM is shown in Figure 3.5.

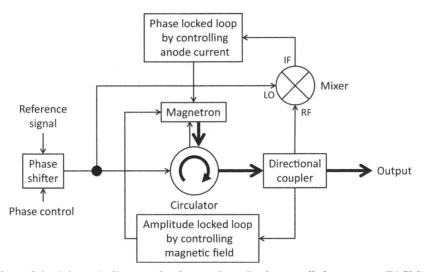

Figure 3.4 Schematic diagram of a phase-and-amplitude-controlled magnetron (PACM).

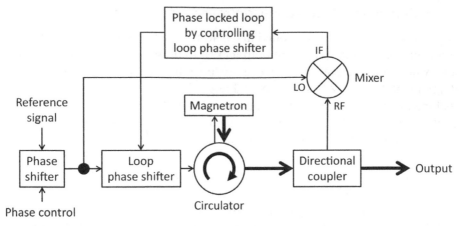

Figure 3.5   Schematic diagram of a power-variable phase-controlled magnetron (PVPCM).

A significant difference between the PVPCM and the PCM is the controller in the PLL system. In the PCM, the magnetron frequency and phase are directly controlled by a feedback loop circuit. In the PVPCM, the reference signal phase is controlled by a loop phase shifter inserted on the way of injection locking signal path. The IF port of the mixer is connected to a control circuit for the loop phase shifter. The magnetron is hence set free from the rigid PLL, and its output power is easily controlled by the anode current.

A major downside of the PVPCM is imperfection of frequency locking. As shown in Figure 3.5, the magnetron frequency locking entrusts only to the injection locking method. If the free-running frequency of the magnetron protrudes outside the locking range expressed in Equation (3.5), this PVPCM concept collapses. For a stable PVPCM operation with a wide range of output power, one should choose a magnetron with small frequency variation with respect to the anode current and magnetron temperature. Otherwise, the injected power of the reference signal should be high enough that the locking range is sufficiently wide relative to the free-running frequency variation of the magnetron.

A 2.45 GHz PVPCM was demonstrated in Japan [26]. The developed PVPCM could change the microwave output from 450 W to 860 W under the injected reference signal power of 4 W. The developed PVPCM can output higher microwave power than the PACM [25]. A fundamental demonstration of a phased array by using two PVPCM was also conducted [26–28]. A 5.8 GHz PVPCM was also developed [29]. A 5.8 GHz CW magnetron seem to be more suitable for assembling a PVPCM than a PCM because its free-running frequency is nearly unchanged by the anode current. The developed 5.8GHz PVPCM could change the microwave output from 160 W to 329 W under the injected reference signal power of 10 W. The phase stability of the 5.8GHz PVPCM was less than $\pm 1°$.

3.2.7 Demonstrations of Microwave Power Transfer by Magnetrons

Magnetrons have been the most commonly used microwave tubes for MPT demonstration experiments since 1960s. The first demonstration system by using the magnetron was developed at Raytheon's Spencer Laboratory in 1963 [14, 30]. In this demonstration, 400 W 2.45 GHz CW microwave power was generated by the magnetron and 100 W of dc power was received. The overall dc to dc efficiency was about 13%. In 1964, an experimental microwave-powered helicopter was demonstrated [31]. A 2.45 GHz magnetron with the output power of 3–5 kW was used for the microwave power source. In 1964, another type of helicopter demonstration by using a microwave beam was conducted [32]. The helicopter received the microwave beam to obtain position information and it could automatically position itself on the beam. A 10.165 GHz magnetron was used as the microwave source in the demonstration.

Regarding MPT to a flying object, a prototype microwave powered airplane for a stationary high altitude relay platform (SHARP) project was demonstrated in Canada [33]. Also an energy transmission toward high altitude long endurance airship experiment (ETHER) was demonstrated in Japan [34]. The transmitting systems of these projects were quite similar. Two feeding ports were equipped with the transmitting antenna to create an orthogonally-polarized dual-polarization microwave beam. As a 5 kW 2.45 GHz magnetron was connected to each port, the total microwave output power was 10 kW. Another feasibility study of MPT from a magnetron transmitter to a prototype airplane for future Mars observation has recently been conducted [26–28].

Magnetrons were also used for terrestrial MPT demonstrations. In 1994 and 1995, 42 m-distance MPT was carried out in Japan [35]. The microwave output was generated from a 5 kW 2.45 GHz magnetron and radiated from a 3 m-diameter parabolic antenna to examine the interconnection method of a rectenna array with 256 rectenna elements. Another demonstration of 40 m-distance terrestrial MPT was demonstrated in Reunion Island, France in 2001 [36]. A 800 W 2.45 GHz magnetron was used as the transmitter. The microwave was radiated from an offset parabolic reflector fed by a pyramidal horn antenna.

With respect to magnetron phased array systems, space power radio transmission systems called SPORTS were developed in Japan [37, 38]. First, a 2.45 GHz phased array with 12 PCMs called SPORTS-2.45 was developed in 2001. Each 2.45 GHz PCM had an output power of more than 340 W with the conversion efficiency of more than 70%, and was connected to either of transmitting antennas: a horn antenna or 8 dipole antennas with a 8-way power divider. The total microwave output reached more than 4 kW. Subsequently, a 5.8 GHz phased array with 9 PCMs called SPORTS-5.8 was developed in 2002. Each 5.8 GHz PCM had an output power of more than 300 W with the conversion efficiency of more than 70%, and was connected to 32 circularly-polarized microstrip antennas with a 32-way power divider. The total microwave output reached more than 1.26 kW. Another phased array system with 2.45 GHz PCMs was developed for MPT from an airship to the ground, which is demonstrated in Japan in 2009 [38]. The transmitter mounted on the airship consisted of two PCMs at 2.46 GHz with the output power of 110 W each. Each PCM was connected to a circularly-polarized radial line slot antenna. The airship was launched above 30 m altitude and the phased array operation was remotely controlled. The microwave power density on the receiving site was fluctuated from 0 to 2.2 mW/cm^2 because the airship including the phased array swayed in the wind during the demonstration.

A magnetron used to be considered as a candidate of the transmitters in some solar power satellite (SPS) concepts. During comprehensive studies on the NASA/DOE SPS reference system [39], Brown discovered that an oven magnetron could be used as a low-cost SPS transmitter with an additional phase locked control loop [40]. He regarded the magnetron as a directional amplifier (the magnetron itself is an oscillator, though), because the magnetron frequency and phase could be synchronized with those of the reference signal by the injection locking method and the PLL [15]. He also developed and experimented radiation cooled magnetrons aiming for space use [41]. A magnetron for the SPS transmitter was also proposed by European Space Agency (ESA), called "Sail Tower" SPS concept [42]. 400,000 magnetrons are arrayed to generate 400 MW power at 2.45 GHz in the Sail Tower.

As a pioneer for MPT in space, a microwave ionosphere nonlinear interaction experiment called MINIX was conducted by Japanese groups in 1983 [37, 38, 43]. This is the world's first successful MPT rocket experiment.

A 2.45 GHz oven magnetron was mounted on the mother transmitter rocket and the microwave power of 830 W was transmitted intermittently to the daughter receiver rocket at the altitude of about 220 km. Nonlinear interactions between a strong microwave radiation and plasma in the ionosphere were measured in the MINIX experiment and studied through computational simulations [44].

3.3 Klystron

A klystron is a linear-beam-type vacuum tube used for the applications of radar, plasma excitation, charged particle accelerators, space communications, and television. In general the klystron is used as an amplifier; whereas a reflex klystron can be used as an oscillator. In this section, only an amplifier-type klystron is described.

3.3.1 Operating Principles

Figure 3.6 shows a cross-sectional view of a two-cavity klystron. The electron beam is emitted from the electron gun section composed of the cathode and the anode, and travels toward the collector. When the electrons pass through the input cavity, the electron velocity is modulated according to the RF electric field. Then in the drift section between the input and output cavities, the electron beam becomes bunched since some electrons are accelerated by the positive RF; whereas some are decelerated by the negative RF.

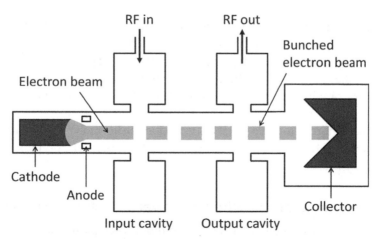

Figure 3.6 Cross-sectional view of a two-cavity klystron.

This means the electron velocity modulation is converted to the electron density modulation. By putting the output cavity to the appropriate position where the electron density modulation is the most extreme, the RF field is strongly coupled to the bunched electron beam and enormously enhanced. In the end, the high-power RF is extracted from the output cavity. The residual electron beam strikes the collector and is converted to heat.

Assuming that the space charge field can be neglected in the gap of input cavity, the electron velocity v after passing through the input cavity is expressed as the following equation,

$$v = v_0 \left(1 + \frac{1}{2} \xi M e^{-j\varphi_0/2} e^{j\omega_k t} \right) \tag{3.7}$$

where v_0, ξ, M, φ_0, ω_k and j is the initial electron velocity, the voltage factor, the coupling coefficient, the nondisturbed transit angle, the angular frequency of the RF input and the imaginary unit ($j^2 = -1$), respectively [11]. v_0 is determined by the dc voltage V_0 in the electron gun section, that is $v_0 = \sqrt{2eV_0/m}$. When the RF input is expressed as $V_R e^{j\omega_k t}$, $\xi = V_R/V_0$. φ_0 is determined by the phase shift when the electron passes through the input cavity gap d with the initial velocity v_0, that is,

$$\varphi_0 = \frac{\omega_k d}{v_0}. \tag{3.8}$$

Assuming $\xi \ll 1$, M is approximated by the following equation,

$$M = \frac{\sin(\varphi_0/2)}{\varphi_0/2}. \tag{3.9}$$

As expressed in Equations (3.7) and (3.9), the degree of electron velocity modulation is dependent on M, which is a well-known Sinc function and monotonically decreases for φ_0 when $\varphi_0 = [0, 2\pi]$ [11]. Finally from Equation (3.8), the gap d should be narrower or the initial velocity v_0 should be faster for obtaining larger M for contributing to highly-efficient klystron operation.

A typical gain of the two-cavity klystron amplifier is 10–15 dB. The gain can be increased by setting more cavities in the drift section. A typical overall efficiency of klystrons is 50–70%.

3.3.2 Demonstrations of Wireless Power Transfer by Klystrons

Klystrons were used as the final stage amplifiers when pioneering the studies of WPT via microwaves. In 1975, one-mile distance WPT was demonstrated by Jet Propulsion Laboratory [45]. In the demonstration, a 450 kW klystron at 2.388 GHz was used as the microwave power source and over 30 kW of dc power was obtained at the receiving site. A part of dc output was used for lighting the lamps to indicate visually how the microwave power density was varied across the receiving array.

Klystrons also used to be considered as the transmitters of the original SPS concept. Glaser, a proponent of the SPS concept, introduced klystrons as the SPS transmitters with efficiency of about 90% at the wavelength of 10 cm [46]. In the NASA/DOE SPS reference system, a 50–70 kW klystron at 2.45 GHz was adopted as the SPS transmitter [39, 47]. More than 100,000 klystrons were assumed in the reference system, since the output power to the terrestrial power grid was set to 5 GW.

3.4 Amplitron

An amplitron is a sort of cross-field vacuum tube devices invented by Brown in 1950s. The name "amplitron" is a Raytheon's trademark, and the device is originally called "platinotron" [48]. The amplitron is classified as a circular format, re-entrant beam, and backward-wave type cross-field amplifier [49]. The forward-wave type platinotron is called "ultron" [11]. The amplitron can be easily modified to an oscillator called "sltabilotron" (also a Raytheon's trademark), by the addition of RF feedback circuit and the application of stabilizing circuit [48]. Amplitrons were used for radar applications and space communications. A remarkable accomplishment of the amplitron was adoption for the S-band power amplifier of the downlink transmitter on the lunar module of the Apollo mission [50]. Also amplitrons used to be considered as the transmitters of MPT systems [30, 51], and the SPS concept as well as klystrons [39].

An amplitron has a coaxial diode structure, which is quite similar to a magnetron. Figure 3.7 shows a cross-sectional view of an amplitron. The cathode in the center plays a role of filament which emits thermal electrons. Multiple vanes in the anode constitutes a slow-wave structure. As the electric field E is applied between the anode and the cathode, the external magnetic field B is applied in the axial direction, as shown in Figure 3.7. Then the electrons move in the azimuth direction by $E \times B$ drift, in the same manner as the magnetron. When the drift velocity of electrons is synchronized with the phase velocity of RF fluctuations excited in the interaction space, the electron sporks emerge and amplify the RF energy. Hence the Buneman-Hartree

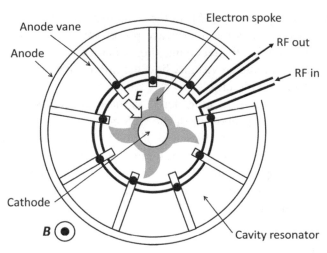

Figure 3.7   Cross-sectional view of an amplitron.

threshold voltage V_{th} expressed in Equation(3.1) is also applicable to the amplitron.

An essential structural difference between the amplitron and the magnetron is the RF ports. The amplitron has both the RF input and output ports matched with the external RF circuits like common microwave amplifiers. In addition, the amplitron has the property of directional amplifier. The RF signal will be neither amplified nor decreased if it is input from the RF output port [48]. One can however exchange the RF input and output ports each other when the external magnetic field direction is reversed.

Some electrons in the amplitron receive plenty of energy from the RF fields and strike the cathode as with the magnetron. This backbombardment energy heats up the cathode unnecessarily or provides secondary electrons. In the case of the amplitron, the secondary electron emission is automatically started by the application of dc voltage and RF input [49]. Therefore, a long-life operation of the amplitron will be expected by using a metal with secondary electron emission capability like platinum as the cathode surface.

As a representative of amplitrons, the QR 1224 was introduced [49, 51, 52]. It could be operated up to 400 kW CW output power with the overall efficiency up to 74%. The 5% bandwidth with nearly constant RF power was obtained at 3 GHz bands with the gain of 8–9 dB. The secondary-emitting platinum cathode was actually used in the QR 1224 for the expectation of longevity.

3.5 Summary

Microwave tube transmitters for MPT applications are described in this chapter. The other types of tubes such as TWT, gyrotron, backward-wave oscillator etc. can be also applied to the MPT system, though they have not been reported for the MPT use yet. For example, a TWT phased array antenna has been reported for satellite broadcasting application [53].

Efficiency is one of the most essential factor in a MPT system. The dc-RF conversion efficiency of the tubes covered in this chapter is around 50–80%, that is comparable to E-class and F-class solid-state amplifiers in the microwave bands these days. Although they can output much higher power than the solid-state devices, microwave tubes require a high-voltage power supply for operation. Moreover, highly-functionalized transmitters such as PCM, PACM and PVPCM include a circulator which suffers additional power loss. In conclusion, the overall efficiency of the transmitter as well as the tubes themselves is a key to an efficient high-power MPT system.

References

[1] Döring, H. (1991). Microwave tube development in Germany from 1920–1945. *Int. J. Electron.* 70, 955–978.

[2] Hull, A. W. (1921). The effect of a uniform magnetic field on the motion of electrons between coaxial cylinders. *Phys. Rev.* 18, 31–62.

[3] Okabe, K. (1929). On the short-wave limit of magnetron oscillations. *Proc. IRE* 17, 652–659.

[4] Hahn, W. C., and Metcalf, G. F. (1939). Velocity-modulated tubes. *Proc. IRE* 27, 106–116.

[5] Varian, R. H., and Varian, S. F. (1939). A high frequency oscillator and amplifier. *J. Appl. Phys.* 10, 321–327.

[6] Copeland, J., and Haeff, A. A. (2015). The true histroy of the traveling wave tubes. *IEEE Spectr.* 52, 38–43.

[7] Kompfner, R. (1947). The traveling-wave tube as amplifier at microwaves. *Proc. IRE* 35, 124–127.

[8] Pierce, J. R. (1947). Theory of the beam-type traveling-wave tubes. *Proc. IRE* 35, 111–123.

[9] Pierce, J. R., and Field, L. M. (1947). Traveling-wave tubes. *Proc. IRE* 35, 108–111.

[10] Sivan, A. (1994). *Microwave Tube Transmitters*. New York, NY: Chapman & Hall.

[11] Tsimring, S. E. (2007). *Electron Beams and Microwave Vacuum Electronics*. New York, NY: John Wiley & Sons.

[12] Kuwahara, N., and Handa, T. (2016). "Development of high power 5.8 GHz cw magnetron–industry's first," in *Proceedings of the 50th Annual Microwave Power Symposium (IMPI 50)*, Orlando, FL.

[13] Collins, G. B. (ed.). (1948). *Microwave Magnetrons*. New York, NY: McGraw-Hill.

[14] Brown, W. C. (1984). The history of power transmission by radio waves. *IEEE Trans. Microw. Theory Tech*. 32, 1230–1242.

[15] Brown, W. C. (1988). The sps transmitter designed around the magnetron directional amplifier. *Space Power*, 7, 37–49.

[16] Mitani, T., Shinohara, N., Matsumoto, H., and Hashimoto, K. (2003). Experimental study on oscillation characteristics of magnetron after turning off filament current. *Electron. Commun. Jpn*. 86, 1–9.

[17] Mitani, T., Shinohara, N., Matsumoto, H., and Hashimoto, K. (2003). Improvement of spurious noises generated from magnetrons driven by DC power supply after turning off filament current. *IEICE Trans. Electron*. E86–C(8), 1556–1563.

[18] Adler, R. (1946). A study of locking phenomena in oscillators. *Proc. IRE* 34, 351–357.

[19] Tahir, I., Dexter, A., and Carter, R. (2005). Noise performance of frequency-and phase-locked cw magnetrons operated as current-controlled oscillators. *IEEE Trans. Electron Devices* 52, 2096–2103.

[20] Tahir, I., Dexter, A., and Carter, R. (2006). Frequency and phase modulation performance of an injection-locked cw magnetron. *IEEE Trans. Electron Devices* 53, 1721–1729.

[21] Shinohara, N., Matsumoto, H., and Hashimoto, K. (2003). Solar power station/satellite (sps) with phase controlled magnetron. *IEICE Trans. Electron*. E86-C(8), 1550–1555.

[22] Hatfield, M. C., Hawkins, J. G., and Brown, W. C. (1998). "Use of a magnetron as a high-gain, phased-locked amplifier in an electronically-steerable phased array for wireless power transmission," in *Proceedings of the IEEE MTT-S International Microwave Symposium (IMS1998) Digest*, Baltimore, MD.

[23] Hatfield, M. C. (1999). *Characterization and Optimization of the Magnetron Directional Amplifier*. Ph.D. thesis, The University of Alaska Fairbanks, Juneau.

[24] Mitani, T., and Shinohara, N. (2007). "Development of a pulse-driven phase-controlled magnetron," in *Proceedings of the 8th International Vacuum Electronics Conference (IVEC 2007)*, Kitakyushu.

[25] Shinohara, N., and Matsumoto, H. (2004). "Phased array technology with phase and amplitude controlled magnetron for microwave power transmission," in *Proceedings of the 4th International Conference on Solar Power from Space (SPS '04)*, Granada, Spain.

[26] Nagahama, A., Mitani, T., Shinohara, N., Tsuji, N., Fukuda, K., Kanari, Y., and Yonemoto, K. (2011). "Study on a microwave power transmitting system for mars observation airplane," in *Proceedings of the IEEE MTT-S International Microwave Workshop Series on Innovative Wireless Power Transmission: Technologies, Systems, and Applications (IMWS-IWPT 2011)*, Uji.

[27] Mitani, T., Iwashimizu, M., Nagahama, A., Shinohara, N., and Yonemoto, K. (2014). "Feasibility study on microwave power transmission to an airplane for future mars observation," in *Proceedings of the 3rd International Conference on Telecommunications and Remote Sensing (ICTRS 2014)*, Luxembourg.

[28] Nagahama, A., Mitani, T., Shinohara, N., Fukuda, K., Hiraoka, K., and Yonemoto, K. (2012). "Auto tracking and power control experiments of a magnetron-based phased array power transmitting system for a mars observation airplane," in *Proceedings of the 2012 IEEE MTT-S International Microwave Workshop Series on Innovative Wireless Power Transmission: Technologies, Systems, and Applications (IMWS-IWPT 2012)*, Kyoto.

[29] Yang, B., Mitani, T., and Shinohara, N. (2017). "Development of a 5.8 GHz power-variable phase-controlled magnetron," in *Proceedings of the 18th International Vacuum Electronics Conference (IVEC 2017)*, London.

[30] Brown, W. C. (1964). Free-space transmission. *IEEE Spectr.* 1, 86–91.

[31] Brown, W. C. (1965). An experimental microwave-powered helicopter. *IEEE Int. Conv. Rep.* 13, 225–235.

[32] Brown, W. C. (1969). Experiments involving a microwave beam to power and position a helicopter. *IEEE Trans. Aerosp. Electron. Syst.* 5, 692–702.

[33] Schlesak, J. J., Alden, A., and Ohno T. (1988). "A microwave powered high altitude platform," in *Proceedings of the 1988 IEEE MTT-S International Microwave Symposium (IMS1988) Digest*, New York, NY.

[34] Fujino, Y., Fujita, M., Kaya, N., Kunimi, S., Ishii, M., Ogihara, N., et al. (1996). "A dual polarization microwave power transmission system for microwave propelled airship experiment," in *Proceedings of the International Symposium on Antenna and Propagation (ISAP '96)*, Chiba.

[35] Shinohara, N., and Matsumoto, H. (1998). Dependence of dc output of a rectenna array on the method of interconnection of its array elements. *Electr. Eng. Jpn.* 125, 9–17.

[36] Celeste, A., Jeanty, P., and Pignolet, G. (2004). Case study in reunion island. *Acta Astronaut.* 54, 253–258.

[37] Matsumoto, H. (2002). Research on solar power satellites and microwave power transmission in Japan. *IEEE Microw. Mag.* 3, 36–45.

[38] Shinohara, N. (2013). Beam control technologies with a high-efficiency phased array for microwave power transmission in Japan. *Proc. IEEE* 101, 1448–1463.

[39] US Department of Energy, National Aeronautics, and Space Administration (1978). Satellite power system concept development and evaluation program, reference system report. Washington, DC: US Department of Energy.

[40] Brown, W. C., and Eugene Eves, E. (1992). Beamed microwave power transmission and its application to space. *IEEE Trans. Microw. Theory Tech.* 40, 1239–1250.

[41] Brown, W. C. (1992). Experimental radiation cooled magnetrons for space use. *Space Power* 11, 27–49.

[42] Seboldt, W., Klimke, M., Leipold, M., and Hanowski, N. (2001). European sail tower sps concept. *Acta Astronaut.* 48, 785–790.

[43] Kaya, N., Matsumoto, H., Miyatake, S., Kimura, I., Nagatomo, M., and Obayashi, T. (1986). Nonlinear interaction of strong microwave beam with the ionosphere minix rocket experiment. *Space Power* 6, 181–186.

[44] Matsumoto, H., and Kimura, T. (1986). Nonlinear excitation of electron cyclotron waves by a monochromatic strong microwave – computer simulation analysis of the minix results. *Space Power* 6, 187–191.

[45] Dickinson, R. M. (1975). *Evaluation of a Microwave High-Power Reception-Conversion Array for Wireless Power Transmission*. NASA Technical Memorandum 33-741. Pasadena, CA: California Institute of Technology.

[46] Glaser, P. E. (1968). Theory of the beam-type traveling-wave tubes. *Science* 162, 857–861.

[47] Glaser, P. E., Davidson, F. P., and Csigi, K. (1998). *Solar Power Satellites*. New York, NY: John Wiley & Sons.

[48] Brown, W. C. (1957). Description and operating characteristics of the platinotron – a new microwave tube device. *Proc. IRE* 45, 1209–1222.

[49] Okress, E. C. (ed.). (1968). *Microwave Power Engineering,* Volume 1. Cambridge, MA: Academic Press.

[50] Grumman Aerospace Corp (1971). *Apollo Operations Handbook Lunar Module LM 10 and Subsequent Volume I Subsystems Data*. New York, NY: Grumman.

[51] Brown, W. C., and Moreno, T. (1964). Microwave power generation. *IEEE Spectr.* 1, 77–81.

[52] Brown, W. C. (1964). Experiments in the transportation of energy by microwave beam. *IEEE Int. Conv. Rep.* 12, 8–17.

[53] Yamagata, K., Tanaka, S., and Shogen, K. (2007). "Broadcasting satellite system using onboard phased array antenna in 21-GHz band" in *Proceedings of the 8th International Vacuum Electronics Conference (IVEC 2007)*, Kitakyushu.

4

Antenna Technologies

Naoki Shinohara

Research Institute for Sustainable Humanosphere, Kyoto University, Japan

Abstract

In this chapter, the antenna and propagation technology for Wireless Power Transfer (WPT) is described. Similar to other wireless applications such as wireless communication and remote sensing, WPT too can be explained by the Maxwell's equation; however, the main requirements – for e.g., the antenna requirements, far field and near field, and beam efficiency – for developing a WPT system differ from those for other wireless applications. In this chapter, the beam-forming technology with a phased array antenna and the technology of direction of arrival by radio waves, i.e., retrodirective technology, are also introduced.

4.1 Introduction

Radio waves are transmitted and received through an antenna, which acts as a converter between the electricity in a circuit and radio wave in space. We do not need a broadband antenna, which is usually required for a wireless communication system, but need a very narrow band antenna because only the radio wave is required as a carrier for the WPT system. Measured wave spectrums for the typical WPT system are shown in Figure 4.1. The radio wave for the WPT system can be modulated to add information to the wireless power, and the broadband antenna can be applied for the modulated wireless power system; however, this is additional technology for WPT.

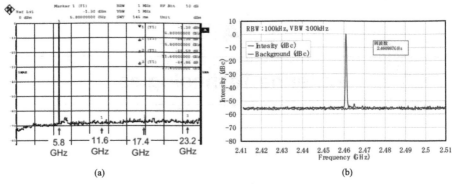

Figure 4.1   Measured Wave Spectrum for a Typical WPT system (a) 5.8-GHz GaN High-Power Amplifier (Mitsubishi Electric and Kyoto University) (b) 2.45-GHz Phase Controlled Magnetron (Kyoto University).

The antenna is an initial condition of Maxwell's equations. Based on these equations, the radio wave is theoretically transmitted omnidirectionally in space from the antenna as a spherical wave. When the distance from the antenna becomes very large, the transmitting sphere wave can be considered as a plane wave. When we assume an aperture antenna of diameter D, a region of the electromagnetic field is classified at the following distance.

$$d > \frac{2D^2}{\lambda} : \text{Far Field (Fraunhofer Region)} \tag{4.1}$$

$$\frac{\lambda}{\pi} < d < \frac{2D^2}{\lambda} : \text{Radiative Near Field (Fresnel Region)} \tag{4.2}$$

$$d < \frac{\lambda}{\pi} : \text{Reactive Near Field} \tag{4.3}$$

We usually use the wireless communication system in the far field. In the far field, the radio wave is considered as a plane wave where, we can always receive equal power wherever a receiving point changes. In the radiative near field, the radio wave cannot be considered as a plane wave; it is a spherical wave, which is derived from Maxwell's equations. We have to consider the power and phase distribution at the plane of each receiving antenna. In the radiative near field and far field, a transmitting antenna and a receiving antenna are not electromagnetically coupled. There is no electromagnetic interference between these antennae. However, in the reactive near field, both antennae are electromagnetically coupled. In the reactive near field, we cannot consider the antennae individually. WPT can be applied for all these distances – the far field, radiative near field, and reactive near field.

One of the most important parameters in the WPT system is the efficiency between the transmitting antenna and the receiving antenna; this parameter is termed "beam efficiency." In the following sections, beam efficiency will be described.

4.2 Beam Efficiency at Far Field

The theory of beam efficiency is well-known and is given as follows:

$$\eta = \frac{P_r}{P_t} = \frac{G_t A_r}{4\pi d^2} = \frac{A_t A_r}{(\lambda d)^2} = \frac{G_t G_r}{\left(\frac{4\pi d}{\lambda}\right)^2} \tag{4.4}$$

where $P_r, P_t, G_r, G_t, A_r, A_t, \lambda, d$ are the received power, transmitted power, and antenna gain of the receiving antenna, antenna gain of the transmitting antenna, aperture area of the receiving antenna, aperture area of the transmitting antenna, wavelength, and the distance between the antennae, respectively. The above equation for beam efficiency at the far field is called the Friis transmission equation.

The antenna gain G and the aperture area of the antenna A are related by the following equation:

$$A = \frac{\lambda^2}{4\pi} G \tag{4.5}$$

The Friis transmission equation can be used for both wireless communication systems and WPT systems. The Friis transmission equation assumes a plane wave under sufficiently far-field conditions in Equation (4.3). Under the assumption of a plane wave, it is easy to increase the beam efficiency many times by increasing the antenna gain. However, please remember that the beam efficiency at the far-field is usually very small, making it easy to drastically increase it.

In the wireless communication system, the parameter $L = \left(\frac{4\pi d}{\lambda}\right)^2$ is commonly called "transmission loss." However, note that L is not a real loss caused by Ohmic loss or dielectric loss during propagation. L is only "loss" at a receiving point caused by the diffusion of the radio wave over the distance. Therefore, if we use a receiving antenna with larger gain, we can receive enough radio waves.

The Friis transmission equation can be used to calculate the receiving power in the far field for energy harvesting or for WPT using diffused radio waves. In the far-field WPT system, wireless power can be easily provided to numerous users simultaneously as that done in a wireless communication system.

4.3 Beam Efficiency at Radiative Near Field

If we need a high-beam-efficiency WPT system instead of a wired system, we cannot use the WPT system at the far field because the beam efficiency in the far field is usually very low according to the Friis transmission equation. We should install a receiving antenna in the radiative near field to increase the beam efficiency of the WPT system. In the radiative near field, we cannot use the Friis transmission equation because the radio wave cannot be assumed as a plane wave. In the radiative near field, the radio wave must be considered as a spherical wave. The experimental data of power density at a receiving antenna aperture at the radiative near field are shown in Figure 4.2 [1]. We cannot assume the plane wave in the experiments. In the experiment, the distance between a transmitting antenna (3 mf diameter parabolic antenna) and a receiving antenna was approximately 42 m. The frequency in the experiment was 2.45-GHz continuous wave. Border between the far field and the radiative

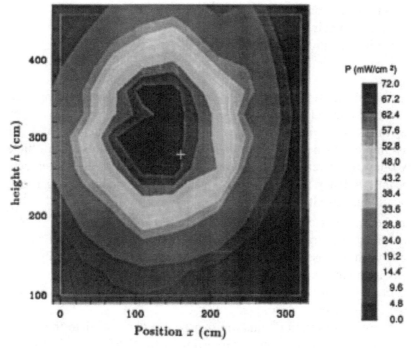

Figure 4.2   Experimental Data of Power Density at the Receiving Antenna Aperture at Distance of 42 m ($\ll$147 m) = $2D^2/\lambda$.

near field $\frac{2D^2}{\lambda}$ was 147m $\gg$ 42m, which means the WPT system was radiative near field system.

Therefore, instead of the Friis transmission equation, we must use τ to calculate beam efficiency η [2–4].

$$\tau^2 = \frac{A_t A_r}{(\lambda d)^2},\tag{4.6}$$

$$\eta = \frac{P_r}{P_t} = 1 - e^{-\tau^2}.\tag{4.7}$$

Small τ indicates a long distance system and far field. Large τ indicates a short distance system and radiative near field. As shown in Figure 4.3, the beam efficiency calculated by Equation (4.4), the Friis transmission equation, and Equation (4.7) with small τ correspond well. This indicates Equation (4.7) can be applied for not only the radiative near field system but also the far field system. With large τ, the beam efficiency obtained by the Friis transmission equation strangely exceeds 100%. This is because of the assumption of a plane wave in the Friis transmission equation. Therefore, for the radiative near field, we must use Equation (4.7) instead of the Friis transmission equation to calculate the beam efficiency.

In Equation (4.7), the amplitude and the phase of a radio wave are assumed uniform at the transmitting antenna. It is also assumed that the

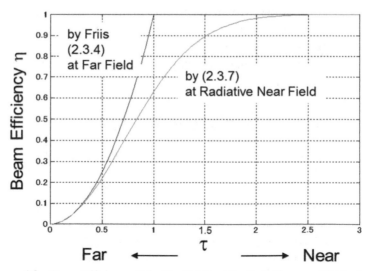

Figure 4.3   Beam Efficiency at the Far-Field and the Radiative near Field using τ.

transmitting antenna and the receiving antenna are at the correct frontal positions.

Based on Equation (4.7), the beam efficiency between the transmitting antenna and the receiving antenna can theoretically reach 100% by adopting the following simple measures:

1. Shorter distance
2. Higher gain antenna (larger aperture antenna)
3. Higher frequency

Another way of increasing the beam efficiency is to use an optimum amplitude taper on a transmitting antenna, such as the Gaussian taper and Taylor taper [4], instead of applying uniform amplitude on the transmitting antenna. When the Gaussian taper is adopted, the beam width increases and side lobes are suppressed, instead of decreasing the front gain. Owing to the wider beam width and suppressed side lobes, the beam efficiency increases. Optimization of the amplitude and the phase on the transmitting antenna is effective for increasing beam efficiency. The phased array is sometimes applied to optimize the amplitude and the phase of the transmitting antenna. Multi-pass transmission is one of the promising technologies that increases beam efficiency. The multi-pass beam forming technology has already been adopted in commercial wireless chargers for mobile phones in the US [1].

In the radiative-near-field WPT system, wireless power is usually transmitted to one receiver with high beam efficiency. When the receiver moves, the wireless beam should reach the receiver to maintain high beam efficiency. A phased array antenna is used to control the beam direction with target detection. If wireless power is provided to multiple users in the radiative near field, the beam-forming technology employing a phased array or the time division method is required.

4.4 Beam Efficiency at Reactive Near Field

The transmitting antenna and the receiving antenna are electromagnetically separate individuals at the far-field and the radiative near field. At the reactive near field distance, however, the antennae are electromagnetically coupled. That is, the antenna impedance and resonance frequency change when the positions of the antennae change. Imagine a conductor and a reflector of the Yagi–Uda antenna. Figure 4.4 indicates the resonance frequency of the transmitting antenna and the receiving antenna in free space and at a distance of 10 cm in the front [5]. The original resonance frequency of 2.45 GHz

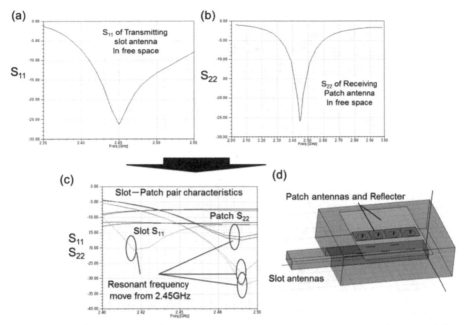

Figure 4.4 (a) S_{11} of the Transmitting Slot Antenna in Free Space, (b) S_{22} of the Receiving Patch Antenna in Free Space, (c) S_{11} of the Transmitting Slot Antenna and S_{22} of the Receiving Patch Antenna at a Distance of 10 cm in the Front, (d) FDTD Simulation Model.

in free space shifts to another frequency when the receiving antenna is located in front of the transmitting antenna. This is because of changes in antenna impedance. This means that the antennae are electromagnetically coupled, and the resonance frequency changes because of interference by the antenna at the front. Because of the changing resonance frequency, the beam efficiency fluctuates at the reactive near field distance, as shown in Figure 4.5. The peaks and troughs of the beam efficiency appear in every half-wavelength.

At the reactive near field, the beam efficiency fluctuates if the antenna impedance is fixed, as shown in Figure 4.5. On the contrary, if the antenna impedance is matched at each distance, the beam efficiency follows the theoretical value obtained using Equation (4.7) [6]. The antenna is a radiator of radio waves at the far-field and reactive near field. However, it is a resonator without radiation of radio waves at the reactive near field. This means that the resonance coupling WPT system can be developed with the antennae for high-frequency waves such as microwaves. The resonance coupling WPT

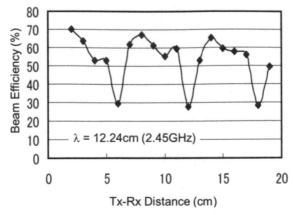

Figure 4.5   Beam Efficiency at the Reactive Near Field by FDTD Simulation.

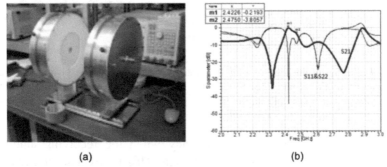

(a)   (b)

Figure 4.6   Resonance Coupling WPT (2.45 GHz) Developed in Japan [7] (a) Developed
Dielectric Resonators, (b) S Parameters at a Distance of 180 mm.

system can be used with not only the typical MHz frequency but also the
GHz microwave frequency. Figure 4.6 indicates a resonance coupling WPT
system developed in Japan. The wireless power is transmitted at 2.45 GHz by
resonance coupling.

In the reactive near field, the antenna becomes a non-radiating resonator.
Resonance coupling WPT can be explained mainly by the inductive coupling theory. There is no clear border between inductive coupling WPT and
resonance coupling WPT. That is, we can explain all WPT technologies
seamlessly, from the inductive coupling WPT to WPT through radio waves.
The only difference between the inductive coupling WPT and WPT through
radio wave is the distance or frequency (see Equations (4.1)–(4.3)). In the
near future, a seamless WPT system will be developed.

4.5 Beam Receiving Efficiency at the Receiving Antenna

The previous sections described the beam efficiency theory. When we choose the optimum parameters of antenna gain, frequency, and distance between the transmitting and receiving antennae, the beam efficiency reaches 100%. To develop a high-efficiency WPT system, we should consider the beam receiving efficiency at the receiving antenna even if 100% beam efficiency is achieved.

An infinite array antenna can theoretically absorb 100% of a transmitted radio wave [8, 9]. In general, the infinite array model can be approximately applied to the analysis of a large array. To achieve 100% beam receiving efficiency, we must match every antenna impedance on the array antenna considering the antenna element spacing and direction of arrival of the radio wave. Please see the detailed theory in the reference [4]. As a theoretical estimation, Figure 4.7 indicates the beam receiving efficiency of the infinite array with circular microstrip antennae whose element spacing is L, and the direction of arrival of the radio wave is θ [10].

The beam receiving efficiency on the infinite array antenna reached 99.9% with the optimum impedance matching condition through finite-difference time-domain (FDTD) simulations [11]. Here, the beam receiving efficiency achieved by the FDTD simulation is shown. Figure 4.8(a) shows a finite

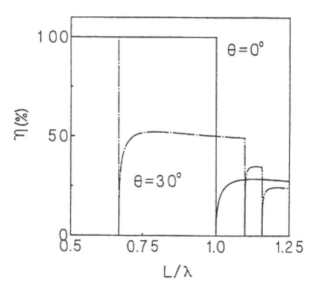

Figure 4.7 Beam Receiving Efficiency with Element Spacing L and Incident Angle θ [10].

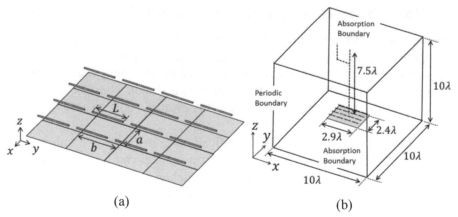

Figure 4.8   (a) Finite 4 × 4 array antenna with parameters L/λ = 0.45, a/λ = 0.6, and b/λ = 0.725 for FDTD simulations and (b) FDTD simulation parameters.

4 × 4 array antenna with parameters L/λ = 0.45, a/λ = 0.6, and b/λ = 0.725. Figure 4.8(b) shows the FDTD simulation parameters of the array antenna. A 2.45-GHz plane wave was input from 10λ above the receiving array antenna.

From the FDTD simulations, the optimum radiation impedance Z of each antenna in the 4 × 4 array antenna was obtained. The results are listed in Table 4.1. The optimum radiation impedance of each antenna in the infinite array antenna was 99.5 − j0.3 (W) [11]. The optimum impedances varied according to the placement of the antenna element.

Figure 4.9 shows the magnetic field obtained through FDTD simulations. The load of Z_L is connected to each antenna. When $Z_L = Z^*$, the highest beam receiving efficiency of the receiving array antenna can be realized. Table 4.2 lists the beam receiving efficiency of each antenna. The efficiency is obtained from the following equation: Efficiency = (power at the load)/(input power

Table 4.1   Optimum radiation impedances

	Re(Z) [Ω]					Im(Z) [Ω]			
	1	2	3	4		1	2	3	4
1	120.3	116.4	116.4	120.3	1	36.7	15.7	15.7	36.7
2	100.1	91.2	91.2	100.1	2	13.9	−3.0	−3.0	13.9
3	100.1	91.2	91.2	100.1	3	13.9	−3.0	−3.0	13.9
4	120.3	116.4	116.4	120.3	4	36.7	15.7	15.7	36.7

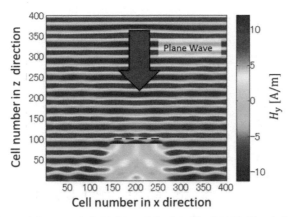

Figure 4.9 Magnetic Field as Calculated by FDTD Simulation.

Table 4.2 Calculated Beam Receiving Efficiency of Each Antenna

	1	2	3	4
1	93.8	83.8	83.8	93.8
2	105.0	98.1	98.1	105.0
3	105.0	98.1	98.1	105.0
4	93.8	83.8	83.8	93.8

at the physical area of each antenna). Efficiency exceeding 100% is obtained because of the difference between the physical area of the antenna and the effective aperture antenna. The antenna at the edge of the array received microwave power from beyond the physical area. The total beam receiving efficiency of the finite 4 × 4 antenna is 95.1%.

The number of antenna elements in the array was then increased, and the beam receiving efficiency of the array antenna was calculated using FDTD simulations. When the antenna with a load Z^* was terminated and the number of antennae was increased, the beam receiving efficiency increased to approximately 100%. As previously mentioned, the beam receiving efficiency of the infinite array antenna is 100% theoretically [8–10] and 99.9% through FDTD simulations [11]. Figure 4.10 shows that the beam receiving efficiency results obtained theoretically and through simulation of the infinite array antenna are in good agreement. When the antenna is terminated with a 50 Ω load, the beam receiving efficiency is very low.

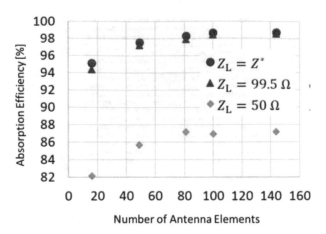

Figure 4.10   Beam Receiving Efficiency of a Finite Array Antenna versus the number of antenna elements.

By combining the beam efficiency theory and the beam receiving efficiency theory, a 100% total wireless efficiency of transmission of radio waves from a transmitting antenna to a receiving antenna in the front can be attained for the WPT system.

4.6 Beam Forming by Using a Phased Array Antenna

When the position of the receiving antenna changes, the radio wave transmitted from the transmitting antenna cannot reach the receiving antenna and the beam efficiency decreases. To maintain high beam efficiency, we must control the beam direction and create an optimum beam form using the phased array antenna technology. We can, of course, control the beam direction by mechanical moving the antenna. However, using the phased array is better than mechanically moving the antenna because of the better speed control, accuracy of beam forming, and life of the system in the case of the former method.

A phased array antenna is a useful technology for electrically controlling the beam direction. The phased array is composed of a number of antennae, as shown in Figure 4.11. We control the phase δ_n and the amplitude a_n of the radio wave transmitted from each antenna using phase shifters or a beam-forming network circuit. The beam direction and beam form is controlled by radio wave interference. The phased array antenna can be applied not only for far-field WPT but also for near field WPT.

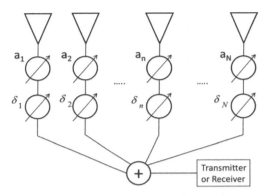

Figure 4.11   Concept of the Phased Array Antenna.

In the case of the far-field, it is easy to calculate beam form $E(\theta, \phi)$ by using the following Equation (4.8), as the product of the element factor $D(\theta, \phi)$ and the array factor $A(\theta, \phi)$. The element factor $D(\theta, \phi)$ is the beam pattern of one antenna element, which is a component of the array antenna. The array factor $A(\theta, \phi)$ collectively indicates the directivity of all antenna elements, considering the phased array as an aperture array.

$$E(\theta, \phi) = D(\theta, \phi)A(\theta, \phi) \tag{4.8}$$

The array factor is determined by the positions, amplitude, and phase of the antenna elements, and it is not related to the types of antennae used. When we consider a one-dimensional, uniformly spaced array of N antenna elements, the array factor is given as follows:

$$A(\theta, \phi) = \sum_{n=1}^{N} a_n e^{j\varphi_n} \tag{4.9}$$

where a_n and ϕ_n are the amplitude and the phase of nth antenna element, respectively. The parameters are shown in Figure 4.3 Equation (4.8) is transformed using Equation (4.8) to obtain the following equation.

$$E(\theta, \phi) = D(\theta, \phi)A(\theta, \phi) = D(\theta, \phi) \cdot \sum_{n=1}^{N} a_n e^{j\varphi_n} \tag{4.10}$$

The phase of an antenna element is attributed to the difference of the antenna position and the phase shifter that can arbitrarily control the phase. The phase of the nth element can be geometrically described as $\phi_n = kd_n \cos \theta + \delta_n$, where k is the wave number; d_n is the nth element spacing; and δ_n is an additional phase shift, for example, caused by a phase shifter. Equation (4.10) can be described using ϕ_n as follows.

$$E(\theta, \phi) = D(\theta, \phi)A(\theta, \phi) = D(\theta, \phi) \cdot \sum_{n=1}^{N} a_n e^{j(kd_n \cos\theta + \delta_n)} \qquad (4.11)$$

The parameters describing a phased array are shown in Figure 4.4. The relations between the beam form of a phased array, element factor, and array factor are shown in Figure 4.12. It is easy to calculate the beam form at the far field using Equation (4.11). In the case of the near field, we should use an additional parameter of the distance from the receiving antenna, which includes the different distances and directions of each antenna element.

We can control an expected beam direction using the phased array. However, unexpected glating lobes, which represent the other beam with identical power as the main beam, form if the element spacing is too large and the beam direction exceeds a value as shown in the following equation.

$$d > \frac{\lambda}{1 + \sin|\theta_s|} \qquad (4.12)$$

where θ_s is the focused direction angle; d is the element spacing; and λ is the wavelength. For example, the unexpected grating lobes occur when $d = 0.75\lambda$ and $\theta_s > 19.5°$, and when $d = 0.6\lambda$ and $\theta_s > 41.8°$. The grating lobes are attributed to the solutions of the sinusoidal function in Equation (4.11). When these lobes are formed, the wireless power is divided among the main lobe and all the grating lobes, and the beam efficiency decreases. These lobes can be suppressed by some methods even if the condition (4.12)

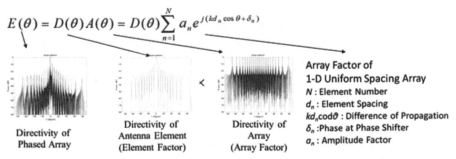

Figure 4.12   Relationships between the beam form of a phased array, element factor, and array factor.

is satisfied [4]. Some technologies are effective for wireless communication and remote sensing. However, they do not consider beam efficiency and only consider the front gain and side lobe level. Therefore, there is no effective technology to suppress the grating lobes in WPT systems.

A phase error, an amplitude error, and problems with the antenna elements can cause a decrease in beam efficiency by causing a shift in the main lobe and increase in the size of the side lobes [4]. It is extremely important to develop a phased array antenna with the errors of the system.

Many phased arrays used in the WPT experiments were developed mainly in Japan after the 1990s (Figure 4.13). The frequencies for the developed phased arrays corresponded to the 2.45-GHz band and 5.8-GHz band microwave. The phased array is usually composed of semiconductors; however, the phased array can be composed of magnetrons, which can be stabilized and can control the phase of the microwaves, in consideration of dc-RF conversion efficiency, microwave power, and development cost.

The first phased array for the WPT field experiment was developed by Kyoto University's group to carry out a microwave-driven airplane experiment named MILAX (microwave lifted airplane experiment) in 1992

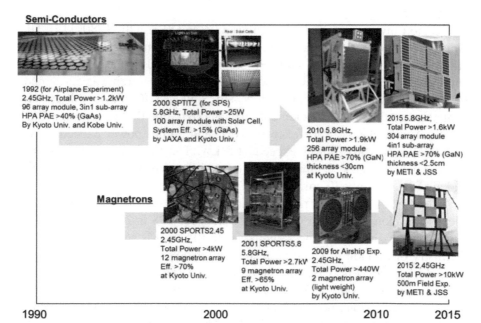

Figure 4.13 History of the Developed Phased Array Antennae for WPT Experiments in Japan.

in Japan. In the phased array, 96 GaAs semiconductor amplifiers and 4-bit digital phase shifters were connected to 288 antenna elements at 2.411 GHz. Thus, there were three antennae in one amplifier sub-array. The diameter of the phased array was approximately 1.3 m. The beam width was approximately 6°. The gain of the amplifier was 42 dB at 0 dBm input. Output power was approximately 42 dBm. The power added efficiency (PAE) of the GaAs amplifier was approximately 40%. The total microwave power was 1.25 kW, and the microwaves consisted of a continuous wave (CW) with no modulation. The developed phased array was re-applied for a WPT rocket experiment, ISY-METS(Microwave Energy Transmission in Space in International Space Year) in 1993.

In FY2000, JAXA (Japan Aerospace Exploration Agency) and Kyoto University developed a phased array as a demonstrator for future Space Solar Power Satellite [12]. It was named SPRITZ (Solar Power Radio Integrated Transmitter '00). The SPRITZ was composed of 100 circular microstrip antennae with 0.75λ element spacing, a 3-bit phase shifter in each antenna, a power divider of 1 to 100, and one high-power GaAs amplifier with an output microwave power exceeding 25 W and efficiency exceeding 15%. Frequency was 5.77 GHz CW, with no modulation. The SPRITZ had solar cells as a dc power source.

Kyoto University started research on WPT and SPS in the 1980s [12]. The University introduced a new research facility, which included a new phased array antenna, for the WPT and SPS studies in FY2010. Based on the GaN and MMIC technology, a new phased array for WPT research was developed. In total, 256 GaN MMIC amplifiers and 5-bit MMIC phase shifters are connected to 256 antenna elements at 5.8 GHz. For the high-power GaN amplifier with class-F, PAE exceeds 70%, and output power exceeds 7 W. The total microwave power exceeds 1.5 kW. The thickness of the phased array, which includes all amplifiers, phased shifters, beam-forming network, and antennae, is only 30 cm.

Based on the phased array technologies in Kyoto University, Mitsubishi Electric developed a thicker GaN phased array for SPS in 2015 (Figure 4.14). It was part of an R&D project dealing with SPS and is supported by METI (Ministry of Economy, Trade and Industry) and conducted by J-space Systems during the period FY2009–FY2014. The thickness of the developed phased array was only 2.5 cm. The phased array was composed of four panels of dimensions 60×60 cm^2 with 76 sub-arrays (3 antennae and 1 amplifier) and weighed less than 1.9 kg. The frequency was 5.8 GHz. The output microwave power from one panel exceeded 450 W and overall, it

Figure 4.14 Phased Array with GaN Developed in 2015 in Japan.

exceeded 1.8 kW. The PAE of the high-power amplifier exceeded 70%. The phased array was applied in a point-to-point WPT field experiment in Japan in 2015, for a distance of more than 55 m.

In addition to these phased arrays, several small phased arrays were developed in Japan. These phased arrays were developed with GaAs or GaN semiconductors. The technologies can be applied not only for WPT but also for remote sensing. Besides, there are phased array antennae with magnetrons that are mainly used in microwave ovens. Kyoto University modified a phase-controlled magnetron with an injection locking technique and a PLL feedback to the voltage source of the magnetron [13], which was originally developed by Brown [14]. Based on the phase-controlled magnetron, the PCM was modified, and a phased array with PCM at 2.45 GHz was developed in FY2000, and one with PCM at 5.8 GHz was developed in FY2001; these arrays are called the Space Power Radio Transmission System for 2.46 GHz (SPORTS-2.45) and Space Power Radio Transmission System for 5.77 GHz (SPORTS-5.8), respectively [15]. SPORTS-2.45 is composed of 12 phase-controlled magnetrons with microwave power exceeding 340 W each and efficiency exceeding 70%. SPORTS-5.8 is composed of 9 phase-controlled magnetrons whose microwave power exceeds 300 W each and efficiency exceeds 70%.

In 2009, Kyoto University developed new phased array with two magnetrons for a field WPT experiment. Here, 2.46 GHz microwave power was

Figure 4.15 Phased Array with Magnetrons Developed in 2015 in Japan.

transmitted with two 110 W output power phase-controlled magnetrons from an airship to the ground [16]. Two radial slot antennae of diameters 72 cm and gain and aperture efficiency of 22.7 dBi and 54.6%, respectively, were used. The element spacing was 116 cm.

In 2015, Mitsubishi Heavy Industries developed a new phased array with phase-controlled magnetrons at 2.45 GHz (Figure 4.15). It is the same R&D project toward SPS supported by METI and conducted by J-space Systems. Eight magnetrons were used in the phased array, and its output microwave power was approximately 10 kW in total. This array was applied for a point-to-point WPT experiment over a distance exceeding 500 m.

4.7 Direction of Arrival

We can control the beam direction by using the phased array antenna and thus maintain high beam efficiency. However, the more important problem is the target detecting technology. If we do not know where a receiver is, we cannot control the beam direction. Various target detection methods, such as those based on the GPS (Global Positioning System), optical technology (laser, CCD cameras, among others), and supersonic technology, are available. In this section, a direction of arrival (DOA) technology employing radio waves is described. When the DOA of the pilot signal from a target can be detected, the power beam can be controlled to the target.

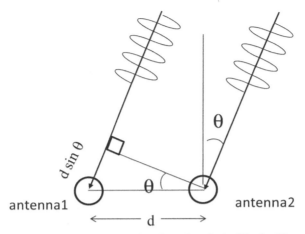

Figure 4.16   Fundamental Direction of Arrival of Radio Waves.

The minimum configuration of the DOA of radio waves is a two-antenna system, as shown in Figure 4.16. The DOA angle θ can be estimated from the phase difference between the two antennae $\Delta\theta$, element spacing d, and wavelength λ as follows:

$$\theta = \sin^{-1}\left[\frac{\lambda \cdot \Delta\theta}{2\pi d}\right](rad).\qquad(4.13)$$

To increase the accuracy of the DOA or to apply multiple signals, numerous DOA algorithms such as those based on the Capon method, linear prediction (LP) method, minimum norm method, MUSIC (multiple signal classification) method, and ESPRIT (estimation of signal parameters via rotational invariance technique), are available. In the WPT system with radio waves, a power beam can be controlled to the target detected by any DOA algorithm. The phased array antenna is preferable for the DOA system, but a mechanical moving antenna too can be applied.

Retrodirective target detection, in which a pilot signal is used for both target detection and antenna positioning detection, is often used for WPT systems. Extensive research is being conducted worldwide on retrodirective arrays for wireless communications. A pilot signal is used in the retrodirective system. As shown in Figure 4.17(a), a two-side corner reflector is the basic retrodirective system. Incoming signals are reflected back in the direction of arrival through multiple reflections off the reflector wall. The Van Atta array is another basic retrodirective system, and it shown in Figure 4.17(b). It is

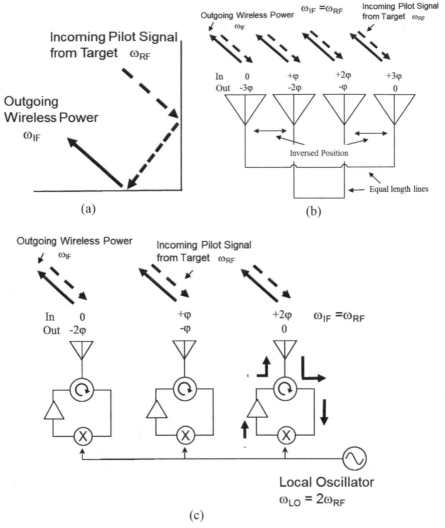

Figure 4.17 (a) Two-sided Corner Reflector, (b) Van Atta Array, (c) Retrodirective Array with Phase-conjugate Circuits.

composed of pairs of antennae equidistantly spaced from the center of the array and connected with equi-length transmission lines. The signal received by an antenna is re-radiated by its pair; thus, the order of re-radiating elements is inverted with respect to the center of the array, achieving proper phasing for retrodirectivity.

Retrodirective systems usually have phase-conjugate circuits in each receiving and transmitting antenna (Figure 4.17(c)); these circuits play the same role as the pairs of antennae that are equidistantly spaced from the center of the array in a Van Atta array. A pilot signal transmitted from the target (amplitude: V_{RF}, angular frequency: ω_{RF}, and phase at each antenna: $+\varphi_n$) is received and re-radiated through the phase-conjugate circuit in the direction of the target. A local signal (amplitude: V_{LO} and angular frequency: ω_{LO}) is used for phase conjugation. Two forms of the pilot signal and the local signal are mixed, and the conjugate signal V_{IF} is obtained, as defined by the following equation. An example of a conjugate signal is shown in Figure 4.18(a).

$$V_{IF} = V_{RF} \cos(\omega_{RF}t + \varphi_n) \cdot V_{LO} \cos(\omega_{LO}t)$$
$$= \frac{1}{2} V_{RF} V_{LO} (\cos[\{\omega_{LO} - \omega_{RF}\}t - \varphi_n] + \cos[\{\omega_{LO} + \omega_{RF}\}t + \varphi_n])$$
$$(4.14)$$

When a low-pass filter is used after mixing, the following signal is obtained. The phase $+\varphi_n$ is changed to $-\varphi_n$

$$V_{IF} = \frac{1}{2} V_{RF} V_{LO} \cos[\{\omega_{LO} - \omega_{RF}\}t - \varphi_n] \qquad (4.15)$$

If we choose $\omega_{LO} = 2\omega_{RF}$, Equation (4.15) becomes

$$V_{IF} = \frac{1}{2} V_{RF} V_{LO} \cos(\omega_{RF}t - \varphi_n) \qquad (4.16)$$

as shown in Figure 4.18(b).

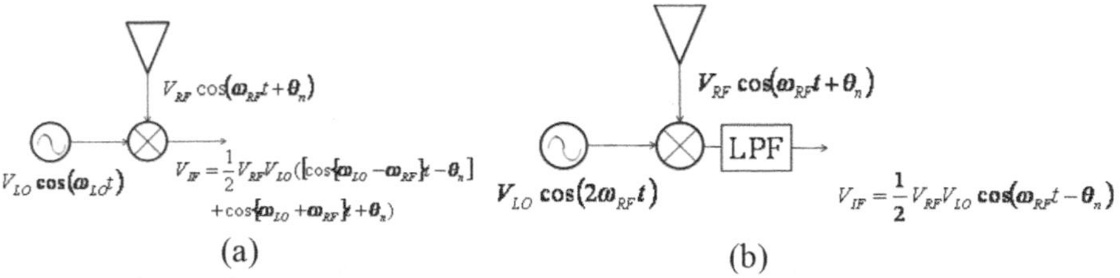

Figure 4.18 Block Diagram of a Retrodirective System with Phase Conjugation: (a) Output without Filtering (b) Output with a Low-Pass Filter (LPF).

The accuracy of the sending direction depends on the stability of the frequency of the pilot signal and the LO signal. To reduce the beam forming error caused by the fluctuation of the LO signal, the same pilot signal and frequency doubler are used instead of the LO signal. Interference between the pilot signal and the wireless power poses a problem if the same frequency is chosen.

Figure 4.19 shows an example of the retrodirective WPT system at 2.45 GHz developed by the Kyoto University and Mitsubishi Electric in 1987. To avoid interferences between the pilot signal and wireless power, two asymmetric pilot signal systems with $\omega_t + \Delta\omega$ and $\omega_t + 2\Delta\omega$ were adopted. Seven antennae were used for receiving two pilot signals and for microwave power transfer.

Various retrodirective systems were used not only for WPT but also for wireless communication. A retrodirective system contains a phase conjugation circuit instead of a phase shifter. Thus, high beam-forming speed is achieved with a low cost. However, the wireless beam is controlled only in the direction of the pilot signal and cannot be controlled in any other directions without the pilot signal.

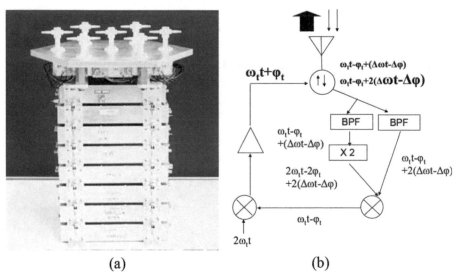

(a)   (b)

Figure 4.19   Retrodirective System with Two Asymmetric Pilot Signals, Developed by Kyoto University and Mitsubishi Electric Corporation (a) Developed Phased Array, (b) Block Diagram of the Retrodirective System with Two Asymmetric Pilot Signals.

References

[1] Hatem, Z., and Saghati, A. (2016). "Remote wireless power transmission system 'Cota'," in *Frontiers of Research and Development of Wireless Power Transfer*, ed. N. Shinohara (Tokyo: CMC Publisher), 185–196.

[2] Goubau, G., and Schwering, F. (1961). On the guided propagation of electromagnetic wave beams. *IRE Trans. Antennas Propagat.* AP-9, 248–256.

[3] Brown, W. C. (1973). Adapting microwave techniques to help solve future energy problems. *IEEE Trans. MTT* MTT-21, 753–763.

[4] Shinohara, N. (2014). *Wireless Power Transfer via Radiowaves (Wave Series)*. Hoboken, NJ: John Wiley & Sons, Inc.

[5] Shinohara, N. (2011). Beam efficiency of wireless power transmission via radio waves from short range to long range. *J. Korean Inst. Electromagn. Eng. Sci.* 10, 224–230.

[6] Chen, Q., Ozawa, K., Yuan, Q., and Sawaya, K. (2012). Antenna characterization for wireless power-transmission system using near-field coupling. *IEEE Antennas Propagat. Mag.* 54, 108–116.

[7] Fujiyama, Y. (2014). *Field Intensity Measurement of the Wireless Electric Power Transmission Equipment using a Dielectric Resonator*. IEICE Technical Report. Beijing: WPT.

[8] Diamond, B. L. (1968). A generalized approach to the analysis of infinite planar array antennas. *Proc. IEEE* 56, 1837–1851.

[9] Stark, L. (1974). Microwave theory of phased array antenna – A review. *Proc. IEEE*, 62, 1661–1701.

[10] Itoh, K., Ohgane, T., and Ogawa, Y. (1986). Rectenna composed of a circular microstrip antenna. *Space Power* 6, 193–198.

[11] Tsukamoto, Y., Matsumuro, T., Tonomura, H., Ishikawa, Y., and Shinohara, N. (2015). "Study on matching condition of an infinite dipole array antenna with reflector for non-leak MPT system," in *Proceedings of the 2015 IEEE Wireless Power Transfer Conference (WPTc2015)*, Boulder, CO.

[12] Matsumoto, H. (2002). Research on solar power station and microwave power transmission in Japan: review and perspectives. *IEEE Microw. Mag.* 14, 36–45.

[13] Shinohara, N., Matsumoto, H., and Hashimoto, K. (2003). Solar power station/satellite (SPS) with phase controlled magnetrons. *IEICE Trans. Electron.* E86-C, 1550–1555.

[14] Brown, W. C. (1988). The SPS transmitter designed around the magnetron directional amplifier. *Space Power* 1, 37–49.

[15] Shinohara, N., Matsumoto, H., and Hashimoto, K. (2004). Phase-controlled magnetron development for SPORTS: space power radio transmission system. *Radio Sci. Bull.* 310, 29–35.

[16] Mitani, T., Yamakawa, H., Shinohara, N., Hashimoto, K., Kawasaki, S., Takahashi, F., et al. (2010). "Demonstration experiment of microwave power and information transmission from an airship," in *Proceedings of the 2nd International Symposium on Radio System and Space Plasma 2010*, Shanghai, 157–160.

5

Efficiency of Rectenna

Simon Hemour, Xiaoqiang Gu and Ke Wu

École Polytechnique de Montréal, Canada

5.1 Introduction

5.1.1 What Is Rectenna?

The rectenna, abbreviated for rectifying antenna, is probably the most important enabling core device for a wireless power transfer (WPT) system via radiofrequency (RF) or electromagnetic waves. It is just the power receiver as opposed to the communication data or radar signal receiver. It begins with the capturing or receiving of a RF energy through the antenna and then converts the received RF energy to a dc or a desired low-frequency power output through a rectifier, which is usually made of a diode detector or a circuit of similar function and its related accessory components. The rectifier integrated with diode in a rectenna is fundamentally the same as the commonly used power or signal detector, even though their design requirements are different. The ideally required RF-dc conversion efficiency for the rectenna is 100% or as close as possible to this "ideal number" since the WPT system is an energy harvester and it is plausibly different from the power or signal detector used in a wireless communication or sensing system whose main role is to extract or detect the signal through the reading or use of the rectified dc output. The rectenna can be used not only for a dedicated WPT from specific RF power sources but also for an energy harvesting from ambient radio waves.

 Figure 5.1 shows the schematic of a typical rectenna which consists of an antenna and rectifier with a diode. Figure 5.2 presents some typical characteristic behaviors of RF-dc conversion efficiency for a rectenna as a function of an input microwave (or RF) power [1]. The RF-dc conversion

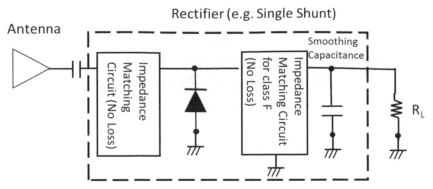

Figure 5.1   Schematic of typical rectenna (rectifying antenna).

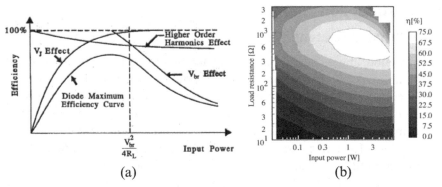

Figure 5.2   Typical characteristics of RF-dc conversion efficiency of rectenna as a function of input RF and microwave power [1] and experimental data of RF-dc conversion efficiency of rectenna as a function of input RF and microwave power and connected load.

efficiency of rectenna changes as the input microwave power or the connected load changes. This power- and load-pulling effect of the RF-dc conversion efficiency is attributed to the nonlinear operation of the diode. The maximum RF-dc conversion efficiency of rectennas reported so far is summarized in Figure 5.3 [2–19] with reference to different operating frequency, designed and developed by various research groups. It can be observed that the conversion efficiency is astonishingly high and it has reached 90% at 2.45 GHz [19] and 80% at 5.8 GHz [17], respectively. For the energy harvesting from ambient radio waves, the RF-dc conversion efficiency is generally very low because the received radio wave power is very weak. It is caused by a junction voltage effect of the diode, which is again related to non-linearity issues.

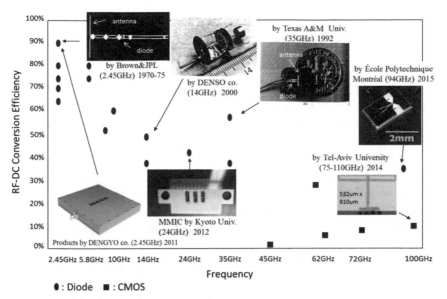

Figure 5.3 Achieved RF-dc conversion efficiency of rectennas reported to date [2–19].

We should consider how to increase the RF-dc conversion efficiency at low input power level, which is the fundamental R&D topic in connection with ambient energy harveting. The frequency-dependent RF-dc conversion efficiency is decided by characteristics of diode [1]. When frequency increases, the efficiency decreases in general.

5.1.2 Rectenna for Energy Harvesting

In this chapter, we focus on RF and microwave-based rectenna for electromagnetic energy harvesting. The electromagnetic spectrum in the form of TV and radio signal, sunlight and countless phenomena is a very rich natural resource like water and air. Electromagnetic energy is omnipresent, some of which is explored for our daily use such as solar power harvested over photon-related electromagnetic spectrum. There is an urgent need for our world energy system to be sustainable from an environmental, economic and social standpoint. However, accomplishing such a sustainability requires a fundamental change in the way that energy is produced and managed. Conventional kilowatt, megawatt and gigawatt power resources coming from centralized power plants have so far dominated our energy production and distribution networks. Unfortunately, they do not have enough potential to improve the energy system to a real sustainable level. Green electricity generation is being implemented on grids as power plant clusters (wind-mill farms, solar power stations...), but it has a very limited impact on the general public because there is a lack of visibility and relevance. The real change would come from our daily life's needs of powering ICT (information and communication technology) devices that operate at a relatively low power usually at mW or μ W-level. They are generally used by individual person or deployed in such distributed systems as future mobile internet of things (IoT) or everything, RFID or wireless sensor networks. In those distributed systems, extremely low power and low duty cycle devices and applications are highly expected with little or without maintenance and human intervention.

Indeed, the new and emerging paradigm proposed by energy harvesting technology is to produce low power electricity anywhere and anytime in a distributive form at device level rather than at town or building level. Its highly-expected potential is in fourfold: (i) It will enable a huge cost-saving and ecological effectiveness on energy transmission (transportation loss, infrastructure installation, maintenance cost, equipment footprint, etc.). (ii) Since it offers mobile energy, it will improve human comfort and reduce human stress in connection with device charging. Ultimately, the impact of this technology is expected to be even higher, for example, (iii) it will yield a great pedagogical value. Users will become more aware of the production & consumption of electrical power, which will make them more willing to save energy (one watt saved is better than a watt produced). It will promote a wide change of mind, a change of personal and societal energy consumption habits toward better use and minimum waste. (iv) Since the source of electricity will be mobile/wearable and produced locally, it will be possible to power and operate devices in areas that are "off the grid", which can be made in an ad hoc and fast deployable manner in case of urgent needs or disastrous situations. Note that 2.6 billion people are still living without reliable power [1].

Renewable energy sources are forms of energy that will remain available for future generations and do not increase the level of carbon dioxide or other pollutants in the earth's atmosphere during energy production. Not only solar, wind, geothermal and ocean energy fits to this requirement but also mechanical energy as movement of body; thermal energy as temperature gradient and radiation energy of electromagnetic fields/waves of the entire electromagnetic spectrum. Thus, this energy can be "harvested" by a mobile electricity "generator" that will provide power for mobile devices.

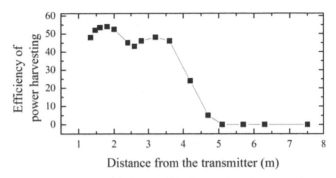

Figure 5.4 Harvesting efficiency of a commercially available device calculated from datasheet, and the Friis equation using 1 Watt of radiated power and 5 dBi receiving and transmitting antennas.

The radio wave energy can be converted to its dc counterpart. This energy can be broadcasted on purpose by power transmitters or stations, but in most cases, ambient radio wave electromagnetic energy is omnipresent, generated by existing base-stations or other wireless services, depending on frequency ranges of interest. The first modern demonstration of such radio energy conversion was shown on television in 1964 by W. Brown and Raytheon Company. They revealed a microwave-powered helicopter that could stay on station at high altitudes for a long time. Even though the wireless power transmission and harvesting research area is more than a century old, as evidenced by the above chronology, changes need to be done to overcome actual low-power limitation of rectifier in connection with conversion efficiency and output generation.

Although a remarkable RF-to-dc conversion efficiency higher than 80% has been reached since about 40 years [19] at high RF power (watt-level), the conversion efficiency of rectifier operating at low-power (μW level) is very low. This is a fundamental hurdle because as the distance increases between transmitter and receiver in a RF power transmission scheme, the received power will decrease with the inverse of the square of the distance, making it more likely to receive low power. Figure 5.4, in which the efficiency of a commercially available power harvester is plotted against distance, shows how μW power **rectifier's efficiency** is of important stakes.

5.1.3 Historical Perspective

Like any new technology, the ambient radio wave scavenging has appeared at the intersection between an application-triggered need and the maturation of a previous technology, that are both interesting to analyze. As semiconductor performance gets improved over years following the Moore's law, the electrical efficiency of computation has doubled roughly every year and a half for more than six decades [20, 21], to a point that as starting from 2000, it has become possible to perform more than 10 computation from less than a micro-Joule (Figure 5.5). This has timely opened a door for the low-duty cycle devices that require a fixed number of computations to perform their tasks, which has in turn triggered the development of endless power generators drawing energy from RF through rectennas. As the historical trends witness it in Figure 5.5, the first rectenna to address efficiently the transformation of propagating radio wave into dc power was at the milliwatt level – thus operating more as a receiver of some remotely beamed power. But the rectenna continued to get improved, enabled by the developments of voltage sensitivity diode devices [22], it has become possible for a rectenna to collect/scavenge/harvest raw RF energy from ambient environments.

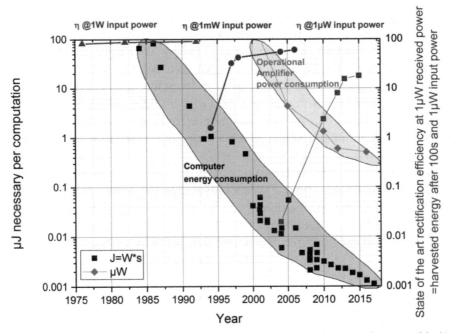

Figure 5.5 Evolution of energy required to operate a computation over the years (black) [20, 21], the energy required to power an Op amplifier (green) [23–28] during one second, both related to the efficiency of RF rectifying technologies for 1W input power (purple), 1mW input power (blue), and 1μW input power (red) [17, 29–37].

5.1.4 The Efficiency Chain

The main objective of the present chapter is to present and highlight the strategies that have been developed to continuously improve the rectenna efficiency by reviewing each of them from the standpoint of the fundamentals of low power EM energy harvesting scenarios (typically −30 dBm and below [38]). Based on this perspective, several points need to be emphasized (i) two-terminal diodes are the mainstream nonlinear devices in low-power rectenna design; (ii) diode's perfect switch model is no more accurate [39]; and (iii) single diode based rectenna is the best choice since more parasitic loss would be introduced by adding more diodes and overall efficiency would be dragged down greatly. The electromagnetic (EM) energy harvester can be seen as a system of chain efficiencies reflecting various steps of the conversion process from the free propagating wave to the terminal dc power as shown in Figure 5.6(a). Note that some of the efficiencies originate from common physical parameters that's may not result in the same effect. As an example, as we will investigate later, we will see that a high junction resistance enhances the rectifier's RF-to-dc conversion efficiency whereas it can adversely quench the matching efficiency. This suggests that there is an optimization or maximization of the overall efficiency in the design process. Figure 5.6(b) provides a graphical insight on the loss mechanism which takes place in each step of rectification process, while (1) translates that into a general efficiency expression:

$$\eta_T = \eta_{rad} \cdot \eta_m \cdot \eta_p \cdot \eta_{RF-dc} \cdot \eta_{SL} \cdot \eta_{dc-dc} \qquad (5.1)$$

5.1.5 Towards Maximum Rectenna Efficiency

Harnessing efficiently the incoming radio wave is about maximizing the aforementioned efficiencies and favor the Carnot Theorem.

- The aforementioned efficiencies can fairly be orthogonalized except for ones that are depending on the junction resistance. Indeed, it will be seen throughout this chapter that a high Junction resistance is harmful for some part of design (antenna impedance, matching circuit) while it is highly valuable for the very core of the rectification mechanism. It will be shown that a junction resistance in the order of thousands of ohm can be considered as a good tradeoff, although it will significantly impact the design.

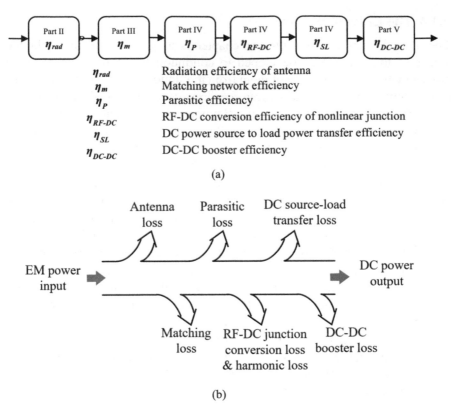

Figure 5.6 Conversion process from free propagating wave to terminal dc power. (a) Efficiency chain and (b) loss mechanism.

- The Carnot's rule is a principle that specifies limits on the maximum efficiency any heat engine can obtain. The efficiency of a Carnot engine depends solely on the difference between the hot and cold temperature reservoirs. Following this principle, it will be demonstrated in the part addressing the rectification mechanism that the zero bias nonlinearity is mainly limited by the thermal voltage as shown in (16), efficiency can thus be linearly increased by increasing the temperature/energy of the RF current flowing through the diode. In other words, input power should be maximized for every ambient energy harvesting scenario.

Keeping those points in mind, the following parts will describe how the efficiency improvement have been addressed recently by the research community.

5.2 Antenna Efficiency

To overcome the impedance discontinuity between the free space which supports the propagation of EM signals and the rectifying diode, the most common way is to introduce two separate components, i.e. an antenna transforming air characteristic impedance (around 377 Ω) to the characteristic impedance of a transmission line and a matching network that matches from the characteristic impedance of the transmission line to rectifying diode's impedance. The antenna is responsible for collecting the most EM power in the free space, thus a high efficiency antenna is desired in rectenna design. Antenna array with a high overall gain can be another efficient way to maximize the collected EM power. Besides, array is possible to be steered to receive the power at certain direction for some applications and provide diversity reception. However, antenna array occupies a much larger space and brings difficulty in some highly desired compact integration design. On the other hand, maximizing the effective antenna aperture is also critical for a given physical size of antenna since this could in turn maximize the power reception. This demands for appropriate choices of antenna technology and topology. The available EM signals exist pervasively in the free space over a wide frequency range such as GSM signals, 3G/4G cellular and Wi-Fi signals, which usually covers from 800 MHz to 2.5 GHz [40]. Broadband or multiband antennas harvesting the energy simultaneously over all these bands can be a highly efficient way to enlarge the injecting power level. When we think from the perspective of an integrated design of antenna and matching network, antenna can be a key component that compensates the matching network loss. Since the diode's impedance is usually very high, designing a high impedance antenna instead of standard 50 Ω can be a solution to make the matching networking design much easier and less lossy. And also, there is a possibility that the matching network can be eliminated by simply using antenna to directly match between the impedance of free space and the diode's high impedance. By doing so, the matching network could be totally removed but such an antenna design can be a challenge. Some recent innovative works done are reviewed below for each individual method.

5.2.1 High Efficiency Antenna

In [41], a high-efficiency broadband antenna is introduced in a relatively low input power scenario (Figure 5.7). This dually-polarized cross-dipole antenna covers a frequency range of 1.8 to 2.5 GHz and has a harmonic rejection property which could further improve the efficiency by rejecting higher harmonics. Compared to rectennas with a similar size, the proposed

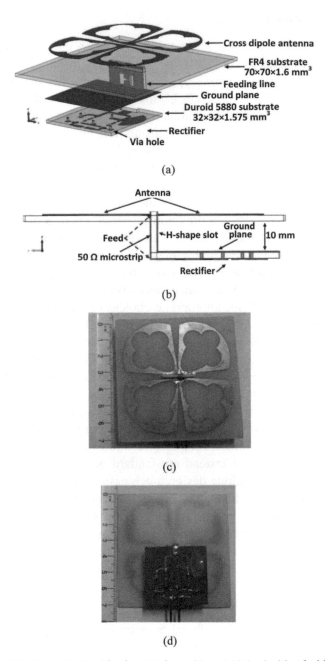

(a)

(b)

(c)

(d)

Figure 5.7   (a) 3-D model, (b) side view, (c) front side and (d) back side of a high-efficiency broadband rectenna [41].

rectenna has a higher output power under the same injecting input power condition. For a wireless power transfer, a narrowband antenna is enough. The efficiency of the antenna and related antenna gain (or aperture area) are important. However, for an energy harvesting, such a broadband antenna is required because the ambient radio waves are modulated. So the broadband, high efficiency, and high gain antenna is suitable in the design of such an energy harvester.

5.2.2 Antenna Array

As another key approach to maximizing the energy reaching diode, a microstrip rectangular patch antenna array is adopted at millimeter-wave frequency [42]. In the near field measurement, the conversion efficiency of around 67% is achieved at 35.7 GHz with a 4 × 4 patch antenna array. A rectenna working at 35 GHz consisting of an antenna array with 16 elements was reported in [43]. The antenna array is formed by 4 × 4 rectangular microstrip patch antennas to fully taking advantage of its high efficiency feature as compared with traditional single element antenna. A 19 dBi absolute gain is achieved by simulation. In order to suppress the second order harmonics, one step-impedance low-pass filter is placed between antenna array and circuitry. By using GaAs Schottky diode MA4E1317, a maximum RF-to-dc conversion efficiency of 67% is obtained when input power is 7 mW. Canadian researchers have proposed a rectenna system including 3-dimentionally folded dipole antennas [44]. The antenna arrays are stacked vertically and connected to channel AC power to a single load. Compared with a single panel with the same size and under the same amount RF illumination, the rectified dc power is enhanced dramatically and reported to be 5 times larger. A 2 × 2 beam-steered phased array antenna was presented and aimed at wireless power harvesting applications [45]. Simply shifting the feeding point position mechanically is able to realize the beam steering without physically rotating the antenna or adding extra active devices. With this feature, the maximum power direction can be tracked in real time. The main beam of the antenna array achieves ±35° in the E-plane and also the antenna array has a high gain of 11.3 dBi and a bandwidth about 13%.

5.2.3 High Impedance Antenna (Better for the Matching)

In order to reduce the loss of matching network between antenna and diode commonly associated with a relative high impedance, high-impedance antennas are introduced in [46]. Based on the design procedures and the developed equivalent circuit, two different shaped folded diploe antennas were demonstrated, namely straight-type and card type. At the operating frequency of 535 MHz (Japanese digital TV broadcasting frequency), antennas have a characteristic impedance of 2000 Ω with the VSWR = 2 criterion. Besides, Japanese researchers have also reported a rectenna with high impedance and high Q antennas [47]. Since a higher voltage injecting into the diode can help increase the RF-to-dc conversion efficiency at relatively low power level, the antennas are designed based on this key point. For example, antenna with impedance of 80 Ω can help increase the conversion efficiency by about 20% when input RF power is 14 dBm as compared with the traditional 50 Ω antenna.

5.2.4 Broadband Antenna

A broadband rectenna covering CDMA, GSM900, GSM1800 and 3G bands was demonstrated to absorb main available RF energy in the air [48]. With input impedance of 300 Ω, the folded dipole antenna is designed together with matching networks to reach a gain of 1.5 dBi over the entire band of interest. Another ultrawideband antenna was designed and tested by British researchers [49]. It is able to exhibit a bandwidth from 450 to 900 MHz which successfully covers DTV, LTE700 and GSM900 bands in the UK. With the help of a hybrid resistance compression technique, such a rectenna can operate in a wide load impedance range from 5 to 80 kΩ with a high conversion efficiency. Measurement results show that an overall efficiency of 42.2% is obtained within an outdoor environment. One fully inkjet-printed broadband planar monopole antenna was reported in [50]. It is able to work from 600 to 1500 MHz which covers several RF legacy signal bands. This antenna can be printed on environment friendly substrates, such as paper and cardboard. Moreover, dielectric coating and silver nanoparticle ink can be realized by a tabletop inkjet printer and metallization can be achieved also on cardboards. Such antenna provides more than 72% radiation efficiency along the band from 800 to 1500 MHz.

5.2.5 Rectenna Integrated Design without Matching Network

Matching network loss is severe in wireless energy harvesting, especially over a low power range. A rectenna design without matching network was

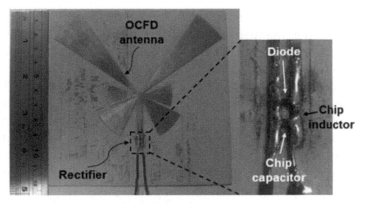

Figure 5.8 Rectenna with a high impedance OCFD antenna and matching network removed [36].

reported in [51]. This class-F rectenna operates at 900 MHz with harmonic terminations to the diode. It is realized on a 0.13 mm PET substrate on which 1μm-thick conductive traces are deposited. Finally, it shows about 48.6% efficiency when the incident power density is only 8 μW/cm². Another research conducted by scholars in Taiwan indicated that an optimal antenna configuration is the structure with an inductively coupled-feed loop instead of conventional patch antenna [52]. By adopting such topology, the radiation efficiency is 82.3% in spite of a lossy substrate. While the radiation efficiency of patch antenna and traditional quasi-Yagi antenna is 29% and 50.9%, respectively. Furthermore, this antenna topology is able to help remove the matching network which is considered as the important energy dissipater in rectenna design. A high impedance antenna was reported by British researchers to successfully eliminate matching network and work in a wide frequency range (Figure 5.8) [53]. This off-center-fed dipole antenna can be tuned to operate in different situations, such as different rectifying diodes and loads. Measurement results demonstrate that such compact and low-cost rectenna is capable of offering a power conversion efficiency of more than 60% over 0.9–1.1 and 1.8–2.5 GHz frequency band, which is suitable for many different applications.

5.2.6 Large Solid Angle High Gain Rectenna

High gain antenna would be desirable to increase with simplicity the rectifying efficiency (thanks to higher input power). Conventional rectenna designs with high gain compact antennas or antenna arrays have already been implemented to improve the received RF power. However, the directive nature of high gain antennas does not fulfill the requirement for rectennas to be omnidirectional. At the same time, dipole rectennas have also been commonly adopted because of their quasi full spatial coverage of linearly polarized radiation. Nevertheless, their low gains do not allow the rectifiers to be efficient at low power operation. In an effort of strengthening the rectifier input power to a higher level, several sources of energy can be collected and rectified by a single element like RF radiations of different frequency bands or different wave polarizations, RF and kinetic energies are efficiently added up to consequently improve the conversion efficiency of diode. Yet, in the last cited scenarios, the improvement is limited to the case where the different sources are collected simultaneously. Attractive strategies have also been proposed like staggered pattern charge collectors (SPCC) where high gain N × N sub-arrays are utilized for increasing the beam width or 3D structures with multiple inkjet-printed antennas covering different directions. These antennas exhibit nonetheless a limited gain, thereby coming up with a limited rectification efficiency. In fact, the combination of RF power conducted by high gain antennas is the most effective way to receive a sufficient power so that it is rectified with high efficiency when low power levels such as ambient reception are involved. That is why cumulating high gain antennas and wide beamwidth radiation are now being explored such as technical schemes described in [54] where a power combining rectenna array with beam-forming matrices was proposed to increase both the received power and the beam width in the E-plane of the rectenna (Figure 5.9).

A high gain antenna array was introduced to help enlarge the capacity of an ambient RF energy harvester in [54]. Five antenna array elements are well-integrated and organized in a pentagon connected with rectifiers using simple passive beam forming networks as shown in Figure 5.9. The operating range of rectifiers has been pushed to a higher input power level, leading to an enhanced efficiency performance. It is reported that the rectification efficiency is improved from 2.6% with a dipole antenna to 14% with the proposed antenna array.

5.3 Matching Network

A matching network is commonly used between the receiving antenna and diodes to cancel the reflections and then maximize the power transfer. At

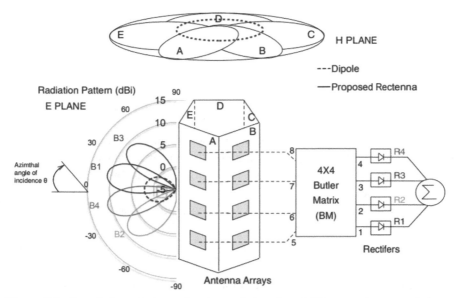

Figure 5.9 Description of a beam forming technique based RF energy harvester. It has a linearly polarized radiation pattern in the E and H plane [54].

the same time, it may introduce an insertion loss and narrow down the operating bandwidth of the rectenna. Since the diode's operating point changes correspondingly with different input power levels, the matching network basically can only operate well within a small power range. In consideration of a low power injecting scenario for the rectifying diodes, we assume that the diode's impedance does not change dramatically. As one of the widely used transmission line in designing the matching network, microstrip line is taken as an example to reveal the loss mechanism. Basically, the losses of a microstrip line consist of three aspects, namely, conductor loss, dielectric loss and radiation loss. The attenuation due to conductor loss could be expressed approximately by [55]:

$$\alpha_c = \frac{R_s}{Z_0 W} Np/m \qquad (5.2)$$

where W and Z_0 are width and characteristic impedance of the microstrip line. $R_s = \sqrt{\omega \mu_0 / 2\sigma}$ is the surface resistivity of the conductor in which μ_0, σ and ω are the free space permeability, conductivity and angular frequency of the microstrip line, respectively.

Since the microstrip line is a quasi-TEM line, the fields around the microstrip line are partly in the air and partly in the dielectric. A filling factor

$\frac{\epsilon_r(\epsilon_e-1)}{\epsilon_e(\epsilon_r-1)}$ is included in the dielectric loss expression [56]:

$$\alpha_d = \frac{k_0\epsilon_r(\epsilon_e-1)\tan\delta}{2\sqrt{\epsilon_e}(\epsilon_r-1)}Np/m \qquad (5.3)$$

where k_0 and $\tan\delta$ are free space wave number and loss tangent of the dielectric.

Microstrip lines radiate at discontinuities as they are not enclosed by a metallic box. The radiation losses include the power radiates into space and radiation carried by surface waves. Radiation loss can be calculated using numerical methods.

The Q factor of a microstrip line could be expressed as below [55]:

$$Q = \frac{\beta}{2\alpha_T} \qquad (5.4)$$

where $\alpha_T = \alpha_c + \alpha_d + \alpha_r$ is the total loss. Then, the insertion loss could be given by [57]:

$$IL = \frac{P_L}{P_{in}} = \frac{P_L}{P_L+P_{diss}} = \frac{1}{1+\frac{P_{diss}}{P_L}} \qquad (5.5)$$

If we relate the insertion loss to the Q factor, we would yield:

$$IL = \frac{1}{1+\frac{Q_m}{Q_r}} \qquad (5.6)$$

Q_m is the Q factor which is reached by design, while Q_r indicates the required Q factor of the matching network as expressed by (5.7). (5.6) indicates if the matching network is designed with a higher Q factor, it will have a lower insertion loss.

$$Q_r = \sqrt{\frac{R_{high}}{R_{low}}-1} \qquad (5.7)$$

in which R_{low} and R_{high} are the lowest and highest impedances in the circuit, respectively. Usually R_{low} is the antenna impedance while R_{high} is the nonlinear junction resistance. It can be clearly observed that higher diode impedance will bring more challenges in designing the matching network.

For a source having 50 Ω impedance, the initial unmatched impedance can be expressed as [56]:

$$|\rho_0|^2 = \frac{\left(R_j + (R_s - 50) \cdot \left(C_j^2 \cdot R_j^2 \cdot \omega^2 + 1\right)\right)^2 + C_j^2 \cdot R_j^4 \cdot \omega^2}{\left(R_j + (R_s + 50) \cdot \left(C_j^2 \cdot R_j^2 \cdot \omega^2 + 1\right)\right)^2 + C_j^2 \cdot R_j^4 \cdot \omega^2} \tag{5.8}$$

where ω is the angular frequency. R_j, R_s and C_j are junction resistance, series resistance and junction capacitance of the diode, respectively. From (5.8), we know that higher frequency will be troublesome. Since R_j is usually much larger than R_s, R_j is a key factor to determine the matching network design for a relatively lower frequency case. In conclusion, designing a matching network is all about to reduce insertion loss, increase operating frequency band and power range.

Bode-Fano criterion [58, 59] is applied to provide a theoretical limit on minimum reflection coefficient magnitude and bandwidth which can be obtained when designing a matching network between arbitrary source and load. The maximum bandwidth can be predicted when given desired reflection coefficient and loading information which is related to the equivalent circuits of diodes in our case. Shockley model indicates that the loading can be approximated as a RC circuit. Thus, the maximum fractional bandwidth based on the Bode-Fano criterion is:

$$\Delta B \leq \frac{\pi}{R_d \cdot 2\pi f_0 \cdot C_d} \cdot \frac{1}{ln\left(\frac{1}{\Gamma}\right)} \tag{5.9}$$

where R_d and C_d are the diode's resistance and capacitance in the equivalent circuit. Γ is the desired reflection coefficient in the band of interest, and f_0 is the targeting frequency. Note that although the Bode-Fano criterion is based on several assumptions, for example, the matching network is lossless and using broadband lossless transformers. And also, the final reflection response is assumed to have only two possible values (ideally equal to Γ in the frequency band, and equal to 1 for any frequencies out of band) which should be realized by introducing infinite elements in the matching network design. All these factors mentioned above make the matching network design with the predicted maximum bandwidth according to the Bode-Fano criterion impossible in practice. However, it still can be an important criterion to set an upper bound for matching network design in reality.

Figure 5.10(a) shows a matching network setup for the efficiency performance comparison of one Schottky diode and one backward tunnel diode.

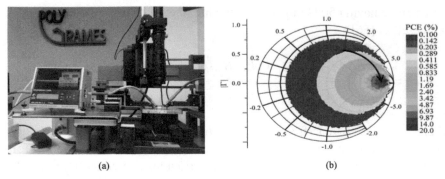

Figure 5.10 (a) Pictures showing part of the setup used as the matching network for comparison of power conversion efficiency of HSMS-2850 and backward tunnel diode. (b) Power conversion efficiency performance in simulation for different matching networks [60].

A probe station and an impedance tuner are used to implement the setup. A simulated distributed efficiency performance on a Smith chart is shown in Figure 5.10(b). It can be clearly observed that a better matching network design will lead to the nonlinear device that completely fulfils its potential to reach a higher conversion efficiency.

In the goal of minimizing matching losses, while operating on high antenna/diode impedance ratio, one has to look for high Q matching network:

- For the application of millimeter wave systems, substrate integrated waveguide (SIW) technology is particularly suitable because of its natural features, such as compact size, high isolation, low radiation and leakage. A novel compact rectenna with a patch antenna and a schottky diode based rectifier circuit integrated inside an SIW resonator was first introduced by A. Collado [61]. However, an inevitable loss was generated due to reflections in the SIW cavity and discontinuity of transmission between patch antenna and rectifier circuit.
- Simply placing high Q SMD inductors in series with the diode would help reduce the VSWR and further reduce the transmission line loss [62]. By adding 3 inductors between transmission line and the diode, the power conversion efficiency rises from 22.5% to 35%. But note that an optimal number of inductors exists as adding more inductors introduce more inductor resistance loss which will reduce transmission line efficiency. In addition, one novel architecture has recently been developed to realize a broadband rectifier which operates at microwave

frequencies without introducing any matching network [63]. With a high impedance inductor, such a rectifier has 40% efficiency in an input power dynamic range from 20 to 25 dBm.

5.3.1 Wide-band Rectifier

A broadband bent triangular omnidirectional antenna covering a frequency band from 850 MHz to 1.94 GHz was demonstrated by M. Arrawatia [64]. This antenna has a VSWR ≤ 2 over the covering band and is designed to receive both horizontal and vertical polarized waves. A voltage of 3.76 V for open circuit and 1.38 V across a load of 4.3 kΩ could be attained when receiving energy from a cell site 25 m away. A broadband rectifier was reported by D. Wang with a relative bandwidth of 57% [65]. With commercial diode HSMS 2820, efficiency performance greater than 50% was obtained when measured from 1.25 to 2.25 GHz. Furthermore, an input power dynamic range greater than 14 dB (12 to 26 dBm) was presented. A compact broadband rectenna based on grounded coplanar waveguide (GCPW) has been demonstrated by M. Nie [66]. The broadband rectifying circuit is composed of a GCPW-fed slot antenna and a voltage doubler principle-based circuit. When a 13 dBm input power is applied, the bandwidth of RF-dc conversion efficiency over 50% was measured to be around 16.3% (2.2–2.6 GHz). While the maximum efficiency is 72.5% at 2.45 GHz.

5.3.2 Rectifiers with a Large Operating Input Range

- Resistance compression network

One approach to overcoming the degradation of system performance with varying incident power level is using transmission line resistance compression networks (TLRCNs). T. W. Barton has derived analytical formulas for TLRCNs and developed both single and multistage rectifiers based on this technique [67]. This scheme helps to maintain a constant RF input impedance over a wide input power range despite of the variation of resistive loads. Operating at 2.45 GHz, a 4 way system has more than 50% conversion efficiency over a variation of input power level of more than 10 dB.

- Load-modulated rectifier for RF micropower harvesting with start-up strategies

Since in a low power harvesting scenario, the harvested energy can normally be hardly applied for any applications in reality due to a low voltage magnitude. A. Costanzio proposed the load-modulated rectifier which is able to offer best sourcing conditions to the dc/dc converter and the start-up stage (Figure 5.11) [68]. This design method is suitable to be integrated with conventional electronic circuits.

- Switching technique

In order to extend input power range of the rectifier but still with an acceptable matching efficiency, researchers in Singapore developed a novel dual-band rectifier [69]. By using a GaAs pHEMT, the voltage on the rectifying diode can be maintained to be almost constant when voltage magnitude exceeds diode's breakdown voltage. Thus, it is reported that more than 30% power conversion efficiency is achievable under input power ranging from −15 to 20 dBm condition. A similar idea was implemented in [70]. One depletion-mode FET is deployed as a switch to compensate the changes of voltage magnitude over the diode. From the measurement, a 40% RF-to-dc conversion efficiency is obtained over a wide power range from −17 to 27 dBm.

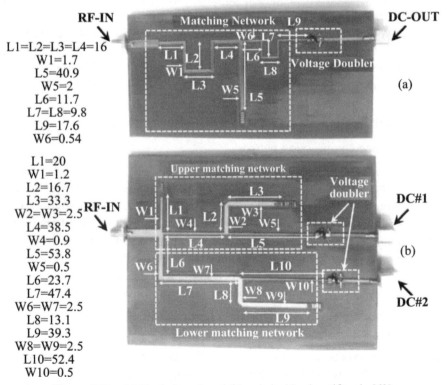

Figure 5.11 (a) Single branch and (b) switched load rectifiers in [68].

- Varactors (Variable reactor)

S. H. Abdelhalem et al. proposed a solution to extend a dynamic range input matching based on varactor diodes [71]. Simply using two varactor diodes to compensate for the variations of impedance due to the change of input power and resistive loads, the authors could obtain a peak efficiency of 60% with $S_{11} < -20$ dB over a more than 12 dB dynamic input power range.

- Branch-line coupler

Recently, a harvester with two identical sub-rectifiers has been designed to realize a high efficient operation within a wide range of input power, operating frequency and loading (Figure 5.12) [72]. A branch-line coupler is inserted between power source and sub-rectifiers. When operating condition varies, the energy reflected by one sub-rectifier has the path to reach the other one. The coupler avoids total energy waste due to input power level, operating frequency or loading information change, although certain energy may dissipate within the coupler. The measured efficiency is able to maintain above 70% when input power changes from 10 dBm to 18.6 dBm.

5.4 Fundamental of the Rectification: RF-to-dc Conversion Efficiency and dc Losses

5.4.1 Conversion Efficiency

At the heart of the RF energy harvester lies a nonlinear junction device where the frequency change takes place. Apart from being converted to the desired dc component, some RF power is distributed among higher-order harmonics

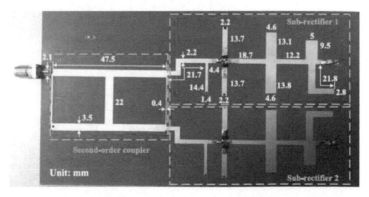

Figure 5.12   Harvester based on a branch-line coupler with two identical sub-rectifiers.

while the rest is dissipated by the junction resistance as a form of Joule heat. As a result, the RF-to-dc conversion efficiency can be defined as the converted dc power divided by the power injecting into the junction resistance which could be written as (5.10a). Since the diode model can be no more considered as a perfect switch in the low power rectifying scenario (0 Ω when forwardly biased and +∞ when reversely biased [1]), another method based on current responsivity $\mathfrak{R}_I$ is introduced in (5.10b), being defined as the ratio of output short-circuit dc current over the injecting power:

$$\eta_{RFDC} = \frac{P_{dc}}{P_{f_0}} \qquad (5.10a)$$

$$\mathfrak{R}_I = \frac{I_{dc}}{P_{f_0}} \qquad (5.10b)$$

The Shockley diode model with packaging parameters is shown in Figure 5.13. The diode's current and voltage (I-V) relationship can be written as:

$$\begin{aligned}
\boldsymbol{I}_{diode} &= I_s \cdot \left[e^{\frac{V_{on}}{n \cdot Vt}} - 1 \right] \\
&= I_s \cdot \left[e^{\frac{v_{f_1} \cos(2\pi f_1 t + \alpha_1) + v_{f_2} \cos(2\pi f_2 t + \alpha_2) - v_{dc}}{n \cdot Vt}} - 1 \right]
\end{aligned} \qquad (5.11)$$

Starting from the Taylor expansion of the diode's I-V relationship around 0 V point (low power case):

$$\begin{aligned}
I(V_j) = I_s \Bigg[&\frac{1}{1!} \frac{V_j}{n \cdot V_T} + \frac{1}{2!} \left(\frac{V_j}{n \cdot V_T} \right)^2 \\
&+ \frac{1}{3!} \left(\frac{V_j}{n \cdot V_T} \right)^3 + \cdots + \frac{1}{k!} \left(\frac{V_j}{n \cdot V_T} \right)^k + \cdots \Bigg]
\end{aligned} \qquad (5.12)$$

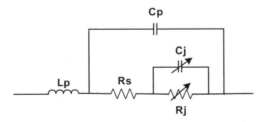

Figure 5.13   Shockley diode model with packaging parameters.

in which V_j is the voltage over the diode, I_s is the saturation current, n is the diode's ideality factor, k is the order of the Taylor expansion and V_T is the diode's thermal voltage which can be expressed by:

$$V_T = \frac{k_B \cdot T}{q} \tag{5.13}$$

where k_B, T and q are Boltzmann constant, operating temperature and electron charge, respectively. Let us assume that the input power has the sinusoidal continuous-wave (CW) waveform with a frequency of f_0, then the voltage can be written as:

$$V_j = V_{f_0} \cdot \sin\left(2\pi f_0 t\right) \tag{5.14}$$

Applying (5.14) into (5.12) and extracting dc current, we have:

$$
I_{dc} = I_s \cdot \left[\frac{1}{2!}\left(\frac{V_{f_0}}{n \cdot V_T}\right)^2 \cdot \frac{1}{2} + \frac{1}{4!}\left(\frac{V_{f_0}}{n \cdot V_T}\right)^4 \cdot \frac{3}{8} \right.
$$
$$
\left. + \cdots + \frac{1}{(2k)!}\left(\frac{V_{f_0}}{n \cdot V_T}\right)^{2k} \cdot \frac{\frac{(2k-1)!}{(k-1)!k!}}{2^{(2k-1)}} + \cdots \right] \tag{5.15}
$$

While the fundamental power can be also calculated based on (5.12) and (5.14):

$$
P_{f_0} = \int_0^{\frac{1}{f_0}} V_j \cdot I\left(V_j\right) dt \cdot f_0 = \frac{I_s \cdot (V_{f_0})^2}{2 \cdot n \cdot V_T} \cdot \left[1 + \frac{1}{8}\left(\frac{V_{f_0}}{n \cdot V_T}\right)^2 \right.
$$
$$
\left. + \frac{1}{192}\left(\frac{V_{f_0}}{n \cdot V_T}\right)^4 + \frac{1}{9216}\left(\frac{V_{f_0}}{n \cdot V_T}\right)^6 + \cdots \right] \tag{5.16}
$$

According to the definition in (5.10b), the current responsivity can be expressed by:

$$
\mathfrak{R}_I(V_{f_0}) = \frac{I_{dc}}{P_{f_0}} = \frac{1}{2 \cdot n \cdot V_T} \cdot \frac{1 + \frac{1}{16}\left(\frac{V_{f_0}}{n \cdot V_T}\right)^2 + \frac{1}{576}\left(\frac{V_{f_0}}{n \cdot V_T}\right)^4 + \cdots}{1 + \frac{1}{8}\left(\frac{V_{f_0}}{n \cdot V_T}\right)^2 + \frac{1}{192}\left(\frac{V_{f_0}}{n \cdot V_T}\right)^4 + \cdots}
$$
$$
= \frac{1}{2 \cdot n \cdot V_T} \cdot \Delta = \mathfrak{R}_{I_0} \cdot \Delta = \frac{1}{2} \cdot \frac{dR_j}{di} \Big/ (R_j)^2 \cdot \Delta \tag{5.17}
$$

where $\mathfrak{R}_{I_0}$ is zero-biased current responsivity and it shows intrinsic characteristic of the rectifying diode, with $R_j = dv/di$ being the differential junction resistance depending on the bias current. Considering the Schottky diode ideality factor n in range of 1 to 2, $\mathfrak{R}_{I_0}$ has a maximum value of 19.34 A/W when T is 300 K, and is solely limited by thermo-ionic transport, opening the door for some interesting applications (see section D – technology, and section F – low temperature applications). The current responsivity $\mathfrak{R}_I$ can be related to voltage responsivity $\mathfrak{R}_V$ with junction resistance R_j [73].

$$\mathfrak{R}_V = \mathfrak{R}_I \cdot R_j \tag{5.18}$$

$\mathfrak{R}_V$ is the ratio of the open circuit dc voltage by the input RF power. By applying Thevenin or Norton equivalent circuits analysis, the total converted current I is divided by loading resistance R_L and video resistance R_v which consists of junction resistance R_j and series R_s. So the current goes through the load resistance that can be expressed as:

$$I_L(V_{f_0}) = \frac{R_j(V_{f_0})}{R_L + R_j(V_{f_0}) + R_s} \cdot I(V_{f_0}) \tag{5.19}$$

Thus the RF-dc conversion efficiency (5.10a) can be concluded as follows:

$$
\begin{aligned}
\eta_{RFDC} &= \frac{P_{dc}}{P_{f_0}} = \frac{(I_L(V_{f_0}))^2 \cdot R_L}{P_{f_0}} \\
&= \frac{P_{f_0} \cdot (\mathfrak{R}_I(V_{f_0}))^2 \cdot (R_j(V_{f_0}))^2}{R_L + R_j(V_{f_0}) + R_s}
\end{aligned}
\tag{5.20}
$$

5.4.2 Parasitic Efficiency

Let us consider again the Shockley diode model. Some injecting power goes through nonlinear junction capacitance and dissipates on series resistance as Joule heat. Parasitic efficiency is defined as the power absorbed by the nonlinear junction resistance over the power injecting into the diode. When one considers a diode without packaging components, the parasitic efficiency can be obtained by simple mathematical treatment using linear circuit basics:

$$\eta_p = \frac{1}{1 + \frac{R_s}{R_j} + (2\pi f_0 \cdot C_j)^2 \cdot R_s \cdot R_j} \tag{5.21}$$

Thus, diodes with larger nonlinear junction capacitance and series resistance will lead to smaller parasitic efficiency.

5.4.3 DC Source to Load Power Transfer Efficiency

Based on (5.19), we know the relation between the current going through the load and the current generated by the nonlinear junction. Thus, the total dc power generated can be calculated by:

$$P_t = (I_L (V_{f_0}))^2 \cdot (R_L + R_j (V_{f_0}) + R_s) \tag{5.22}$$

While the dc power reaching the load is:

$$P_{dc} = (I_L (V_{f_0}))^2 \cdot R_L \tag{5.23}$$

Then the dc source to load transfer efficiency is defined by the ratio of the dc output power reaching the load over the total rectified dc power:

$$\eta_{SL} = \frac{P_{dc}}{P_t} = \frac{R_L}{R_L + R_j (V_{f_0}) + R_s} \tag{5.24}$$

Although junction resistance $R_j (V_{f_0})$ varies with the change of input voltage magnitude, it can be approximately considered to be zero-biased junction R_{j0} resistance in a low power harvesting case. The same assumption is made to current responsivity which means $\Re_I (V_{f_0}) \approx \Re_{I0}$ within the low power range of interest. If we combine both RF-to-dc conversion efficiency (5.20) and dc source to load transfer efficiency (5.24) together [39], we can get:

$$\eta_{SL} \cdot \eta_{RFDC} = \frac{R_L}{R_L + R_{j0} + R_s} \cdot \frac{P_{f_0} \cdot \Re_{I0}^2 \cdot (R_{j0})^2}{R_L + R_{j0} + R_s} \tag{5.25}$$

According to (5.25), optimum load information can be obtained. Making the first derivation of (5.25) as a function of load resistance to 0:

$$\frac{P_{f_0} \cdot \Re_{I0}^2 \cdot (R_{j0})^2 \cdot (R_{j0} + R_s - R_L)}{(R_{j0} + R_s + R_L)^3} = 0 \tag{5.26}$$

Since $P_{f_0} \cdot \Re_{I0}^2 \cdot (R_{j0})^2$ will not equal 0, the only solution is:

$$R_L = R_{j0} + R_s \tag{5.27}$$

In order to see whether this value will lead to the maximum or minimum efficiency, let us do a second derivation of (5.25):

$$\frac{d^2 (\eta_{SL} \cdot \eta_{RFDC})}{dR_L^2} = \frac{6 \cdot P_{f_0} \cdot \Re_{I0}^2 \cdot (R_{j0})^2 \cdot R_L}{(R_{j0} + R_s + R_L)^4} - \frac{4 \cdot P_{f_0} \cdot \Re_{I0}^2 \cdot (R_{j0})^2}{(R_{j0} + R_s + R_L)^3} \tag{5.28}$$

When substituting R_L with $R_{j0}+R_s$, $\frac{d^2(\eta_{SL}\cdot\eta_{RFDC})}{dR_L^2} < 0$ which means $\eta_{SL}\cdot\eta_{RFDC}$ would reach a maximum value. Thus in an extremely low power range, the optimum load resistance is considered to be equaling zero-biased junction resistance R_{j0} since usually $R_{j0} \gg R_s$. Based on the considerations aforementioned, (5.25) can now be simplified as:

$$\eta_{SL}\cdot\eta_{RFDC} = \left(\frac{\mathfrak{R}_{I_0}}{2}\right)^2\cdot R_{j0}\cdot P_{f_0} \qquad (5.29)$$

Equation (5.29) is a major result providing several effective methods to increase harvesting efficiency:

- By deploying diodes or other nonlinear devices with larger zero-bias current responsivity $\mathfrak{R}_{I_0}$. This can be understood as improving the ratio resistance curvature over resistance value, where the curvature (nonlinearity) fuels the RF-to-dc efficiency and the non-variating part of the junction resistance leads to Joules losses both at RF and at dc.
- By using diodes with larger junction resistance and lower series resistance. The role of the value of junction resistance should be handled with care depending on the efficiency chain standpoint. It has to be seen here as a way to enhance the voltage through the I(V) curve of the diode through the Ohm's law, which in turn will expose the incoming signal to the nonlinearities of the diode.
- By enlarging the input (collected) power P_{f_0}, the operating point can be shifted to another region with a higher conversion efficiency as it can be instinctively deduced from Carnot's theorem (with relation to the second law of thermodynamics).

5.4.4 Enhanced Nonlinearity

- *Alternative Technology based on tunneling transport*

As it has been demonstrated in the previous section, Schottky diode zero-bias current responsivity is limited by thermal voltage. But it is not the case for the technologies that rely on different transport mechanisms. For instance, an approach using herterostructured backward tunnel diodes in rectifier design has recently broken the record of low power rectification efficiency [60]. This has been made possible thanks to an AlSb high barrier height blocking the thermo-ionic transport, while allowing conduction through an energy window for tunneling (Figure 5.14).

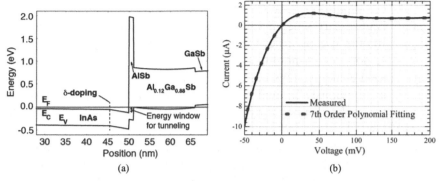

Figure 5.14 (a) The calculated energy band and (b) measured I-V curve and its polynomial fitting of the backward tunnel diode used in [60] zero bias current sensitivity of 23.4A/W.

Two other technologies, namely MIM diodes and Spindiodes should soon give birth to stronger nonlinearities thanks to their transport mechanisms, but they are not yet matured enough to outperform the Schottky standards. As of 2017, the state of the art in metal-insulator-metal (MIM) diode with reasonable junction resistance (18kOhm) is 4.4A/W [74]. This should be compared to the maximum Schottky current sensitivity of 19.4A/W at ambient temperature.

Spindiode technology, where the transport mechanism spin-dependent shall also be a promising technology, as it can take advantage of both charge and spin of the electron compared to those which only use the electron charge [22].

- *Using reactive nonlinear devices*

The bottom line is the diodes and resistance used to convert/rectify signal are, first and foremost, intrinsically resistive, which means lossy. Their operation is based on a nonlinearity of the real part of their impedance. That is certainly welcomed by the circuit designer for the ease of the matching circuit, but this leads inexorably to Joules losses. This reasoning has recently lead to the interesting proposition of building a RF energy harvester around a nonlinear reactance device [75]. Two type of varactors can be used for this purpose: a saturable-core reactors (inductance) or a reverse-biased-controlled semiconductor depletion zone thickness (junction capacitance), the latter being more mature at microwave frequencies. The energy harvester circuit describes a two-stage circuit using (i) a nonlinear resonator to basically down-convert the RF signal to a very low frequency, and (ii) a modified boost converter, with balanced input, to operate with positive and negative inputs. This second stage is necessary since a varactor does not operate at zero frequency. Conceptually, the use of a nonlinear reactance may solve the problem of any possible not-enough-nonlinear devices (i.e. any diode operating at low power), since the energy stays in the nonlinear resonator until it is converted. In reality, it is obviously limited by the parasitic resistance, although nowadays varactor diodes can be fabricated with very low resistance (in the order of 0.1Ω [75]).

The problem of such a kind of high Q circuit is that matching circuit losses play a critical role in the quality factor. It is also too sensitive to source and load variation. The final drawback is that since the output frequency has to be low enough to be exploitable by a power electronic boost converter, the two input microwave signals have to be separated by 1 kHz each, which is technically unachievable. Duplexer of 50 MHz can be used, but leads to a lower quality factor of the nonlinear resonator and higher operating frequency of the boost converter, where it is itself not at its best efficiency.

5.4.5 Increasing Junction Resistance

Another way of looking at the response of a nonlinear device is to measure the current sensitivity (short-circuit output current per unit of input power, (5.10b)) since it will linearly affect the efficiency (5.29). Some commercial diode voltage sensitivities are plotted in the next figure as a function of junction resistance. It can be understood that all the state of the art diodes have very similar current responsivity since they are almost aligned to the maximum nonlinearity (purple line). MIM junction and MTJ (magnetic junction are good candidates for validating this approach. An example is given in Figure 5.15, where a set of samples with different junction resistance were built by simply changing the section of the junction. Increasing the junction resistance thus increases the apparent nonlinearity, which is highly beneficial to the rectification but is limited by the capabilities of the matching circuit through its quality factor.

5.4.6 Low Temperature Operation

As have been mentioned before, thermionic emission is the main feature of Schottky diodes, thus the zero-bias current responsivity $\mathfrak{R}_{I_0}$ has a maximum value. However, $\mathfrak{R}_{I_0}$ is inversely proportional to the operating temperature, this gives us a simply way to enhance Schottky diodes' nonlinearity by

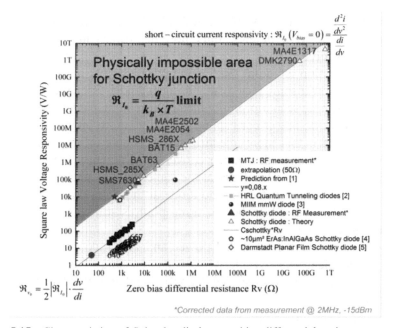

Figure 5.15 Characteristics of Schottky diode, zero bias differential resistance vs square law voltage responsivity. Higher zero bias resistance results in higher dc voltage across the junction (but might be more difficult to match). The same responsivity-junction resistance relationship is observed for different technologies (Schottky diodes, tunnel diodes, MIM diodes and Spindiodes).

decreasing the operating temperature. For other types of diode, low operating temperature sometimes is also able to bring advantages. It is reported that the differential gain could increase from 1000 V/W to about 4500 V/W when one commercial tunnel diode (Model Herotek DT8012) is operating at 4 K under input power condition of –30 dBm to –20 dBm [76]. This diode is widely used in low-level microwave power detectors and thus theoretically can be implemented in low power rectifier circuit. Of course, the cost in connection with the cooling system may be prohibitive in many commercial applications, but this solution shall make sense in the temperature of outer space.

5.4.7 Enhance Input Power

Besides the method of collecting more energy from single power source, introducing other different types of power sources can be an attractive alternative. When one more dc power source is accessible, it can be the bias for the other AC power source. DC bias could shift the operating point to a region with higher conversion efficiency, thus more dc power can be rectified from the AC power source.

Another situation is adding one or more AC power sources, let us take hybrid two AC power source harvesting as an example. For a rectenna, hybrid AC power harvesting scenario simply means more input power and will result in a larger output dc power correspondingly. One more interesting point for hybrid harvesting is that more injecting power is able to drag the operating point to another region with a higher conversion efficiency. Since the hybrid harvesting captures two advantages at the same time, theoretically speaking, 2 units input power injecting into the diode at the same time will lead to 4 times dc output power when the operating point is shifted to a region with two times of efficiency.

When we compare hybrid AC power and AC + dc power harvesting topology, it is clear that hybrid AC power is capable of offering more dc power output since dc power source is applied and consumed as the bias for the diode.

- AC power combination

A hybrid harvester capable of collecting both RF radiation and mechanical vibration power was introduced by Polytechnique Montreal [78]. This work

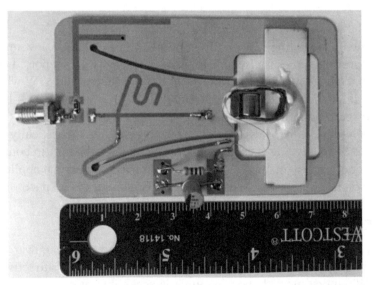

Figure 5.16 Prototype of a hybrid energy harvester collecting RF and kinetic power simultaneously [77].

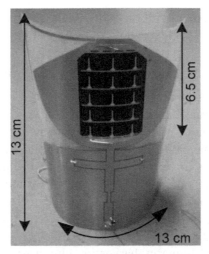

Figure 5.17 Dual band hybrid EM and solar harvester based on a flexible substrate [79].

proved that a combination of uncorrelated RF and vibration power into a Schottky diode would dramatically improve the RF-to-dc efficiency. The experiments show that up to 6 dB gain can be achieved by this hybrid harvester compared to the single source harvesting. Later, in the same group, an integrated design of RF and mechanical energy harvester is proposed. A permanent magnet and coils based mechanical generator and an F-shaped antenna is all combined on a single substrate, with a similar size of credit card. This integrated cooperative harvester has a strong resilience in an ever-changing environment and more suitable for real-world application [77].

• dc + AC power hybrid harvesting

Besides mechanical vibration power, a solar electromagnetic rectenna was demonstrated by F. Giuppi et al. from Spain [79]. Based on a flexible substrate, both solar cells and an antenna are nicely placed although solar cells could directly obtain dc current without any rectifier. Thermal energy coming from power amplifier is also mentioned to be a potential recycling source by adopting a thermoelectric material in the same paper.

Recently, a hybrid electromagnetic and solar energy harvester with communication functions, operating at 2.4 GHz ISM band, was proposed [80]. When RF input power is only −12.6 and −15.6 dBm, it is able to offer enough voltage and energy to a power management unit (PMU) bp25504 operating at "cold start" and "hot start" status, respectively. The required RF energy can be reduced by 50% compared with the individual RF harvester, when utilizing PMU with such hybrid harvester under the room light irradiation condition of 334 lx.

- Multisine or high peak to average power ratio (PAPR) signals

Spanish scholars showed that an improvement of RF-dc efficiency could also benefit from pre-generated signal waveforms [81]. After testing several kinds of signal waveforms, they found that high peak to average power ratio (PAPR) signals contribute to increase RF-dc conversion efficiency.

Further researchers in Portugal developed two schemes to address the issue of amplifying large multi-sine PAPR signals simply by adopting a spatial power combination [82]. One highlighted design is to use synchronization in mode-locked oscillators to set a phase reference for high PAPR multi-sine waveforms which enables to maximize the RF-dc conversion efficiency.

Recently, another RF energy harvesting circuit was developed to collect time-varying signals such as multi-tone or digitally modulated signals with random modulations [83]. It is demonstrated that a time-varying envelop signal can lead to a higher efficiency than conventional continuous wave signals.

5.4.8 Synchronous Switching Rectifiers (Self-Synchronous Rectifier)

By using a self-synchronous and self-biased E-pHEMT rectifier, an extremely high efficiency peak of 88% was obtained at 16 dBm by University of Cantabria [84]. With a self-resonant drain terminating coil, this approach is used in 900 MHz wireless powering links. M. Litchfield has shown that GaN X-Band power amplifiers could operate as self-synchronous rectifiers [85]. The self-synchronous technique is realized by the finite gate-to-drain non-linear capacitance in the intrinsic GaN HEMT providing as a feedback and further enabling RF power coming out of the gate port. A RF-dc efficiency of 64% was obtained when the MMICs work in rectifier mode. An Enhancement-mode Pseudomorphic High Electron Mobility Transistor (E-pHEMT) based self-biased and self-synchronous class E rectifier was proposed by M.N. Ruiz in Spain [86]. The self-synchronous feature is again realized by the device gate-to-drain coupling capacitance, further shrinking the size of the design. And gate-to-source Schottky junction makes self-biasing possible to enhance the efficiency. A high-efficiency zero-voltage-switching (ZVS) AC-dc light-emitting-diode (LED) driver was presented by J.-W. Yang [87]. By using a self-synchronous rectifier instead of an output

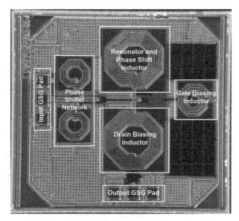

Figure 5.18 2.4 GHz CMOS class-E synchronous rectifier [88].

diode, the conduction loss is reduced dramatically and circuit schematic is simplified thus reducing the cost as no additional circuit is needed due to its self-driven feature. S. Dehghani proposed a class-E rectifier using 0.13 um CMOS technology including an input matching and a self-biased gate (Figure 5.16) [88]. Matching is implemented on chip and this rectifier shows a peak RF-to-dc conversion efficiency of 30% at 2.4 GHz. One highly sensitive RF energy harvester consisting of a synchronous CMOS rectifier and a small loop antenna was reported by M. Stoopman [89]. One 5-stage synchronous rectifier with a complementary MOS diode is well-designed to enhance the harvester's capacity to store energy for a longer time. When measured at 868 MHz, an end-to-end maximum PCE of 40% and a sensitivity of –27 dBm to generate 1V across a capacitive load were attained.

5.4.9 Harmonics Management

Harmonic generation and intermodulation product are a function of input power and rectifying device nonlinearities. Although the amount of power available in typical ambient energy harvesting application is not enough to drive the rectifying device in a strong nonlinear behavior, a harmonic management is mandatory above 10 mW.

- Harmonic termination

A typical way to handle harmonics in Power Amplifiers/rectifiers is to terminate it so to maximize the output power at the frequency of interest (in the present case, dc). Specific harmonic terminations at the virtual drain reference plane in the transistor can be used to reach specific class of operation [90]. One concurrent 2.45 GHz and 5.8 GHz rectifier with harmonic termination was proposed [91]. A dual band matching network is designed and a quarter-wave length open stub of 8.25 GHz which is the mixing component of the two-tone signal is connected with the diode's cathode as the harmonic termination. And a spurline notch filter is acting as both 2.45 GHz and 5.8 GHz band stop filter to maximize the RF-to-dc power conversion efficiency. Measurement results show that 64.8%, 62.2% and 67.9% at 2.45 GHz, 5.8 GHz and two-tone injecting scenario are achieved, respectively.

Of course as it is the case with matching network, quality factor is of high importance to reinject the totality of the harmonics energy to the rectifying device.

The group of Popovic has adopted a Gallium Nitride (GaN) HEMT to rectify microwaves. They proposed and developed a rectenna with a GaN HEMT which is the same as class inverse-F amplifier [92]. Any amplifier will function as a rectifier, and at microwave frequencies as a self-synchronous rectifier without any gate drive. They calls it 'PA-rectifier' and achieves 85% efficiency at 2.11 GHz, 8–10W input. It can be used as class inverse-F amplifier whose PAE (Power Added Efficiency) is 83% at 2.11 GHz, 8W input.

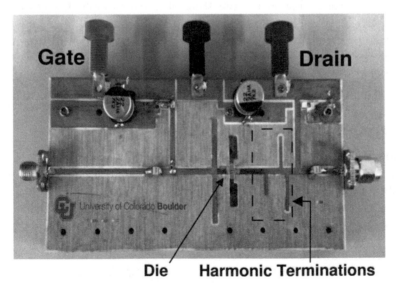

Figure 5.19   2.14 GHz GaN HEMT rectifier based on a class-F^{-1} power amplifier [92].

- Harmonics recycling

Recently, D. Allane et al. have demonstrated an augmented tag which implements the energy harvesting of wasted third harmonics generated by passive RFID UHF chips [93]. The UHF RFID chip was modeled by a diode-based voltage multiplier circuit. A three-port matching network works to redirect fundamental signals between RFID and reader sections and third harmonics signals to the harmonic harvesting section. Finally, the RFID tag could implement communication with reader and powering an external temperature sensor simultaneously. When a 10-dBm input power applied, a dc power of 39 uW was obtained from the third harmonic product of a RFID chip. S. Ladan et al. reported a full-wave rectifier working at 35 GHz including harmonic harvesting features [94]. By optimizing the harmonic harvesting rectifier, the efficiency has been increased by 11% from 23% to 34% compared to the previous conventional voltage doubler rectifiers at 20 mW RF input power level.

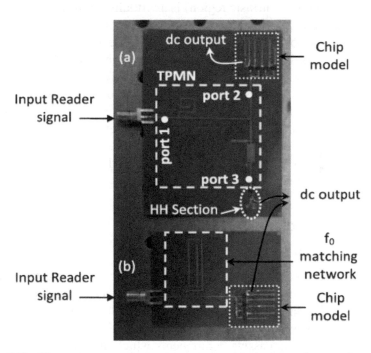

Figure 5.20 Harmonic power harvesting system. (a) Augmented tag using the chip model in (b) [93].

5.4.10 Low Transistor Conduction Losses

Improved rectifying efficiency has also been reported using transistor. Among other technologies, GaN based transistor offers a very high switching current density combined with high voltage operation. This means that when operating at high power, the conduction loss to rectified power ratio generally surpass circuit that utilizes Gallium Arsenide (GaAs) devices.

5.4.11 Diodes with Low Nonlinear Junction Capacitance

VDI zero biased diode based rectifier was reported to have the highest efficiency within the power range from –40 dBm to –20 dBm compared to those based on other Schottky diodes [39]. Since parasitic loss plays a significant role in low power harvesting [22], VDI zero biased diode based rectifier provides superior performance due to its low nonlinear junction capacitance. The extracted total capacitance is reported to be only 25 fF. While by contrast, PIN diodes have three regions, namely P-region, I-region and N-region. I-region (intrinsic region) is a virtually undoped intrinsic layer which separates both heavily doped anode (P-region) and cathode (N-region). Due to the existence of the intrinsic region, PIN diode is able to tolerate a high voltage when reversely biased and is suitable for high voltage/power rectifier design. Using spindiodes would be another alternative to minimize parasitic losses [22]. The external bias dependent resistance of such diodes could not only benefit the matching network design, but also reduce parasitic loss. For example, the extracted junction capacitance and series resistance are only 10 fF and 1 Ω, respectively for a 900 Ω spindiode. Another factor helping lower the loss is that the junction capacitance could be reduced and further improving the frequency performance.

5.5 Booster Efficiency

5.5.1 Commercial Circuits

In order to power post-circuitry using rectified dc current, a boost dc-dc converter is necessary to lift the magnitude of the voltage to meet the requirement under different input voltage levels with a good efficiency. Also high frequency dc-dc converter is able to shrink the size of the passive and reduce the value of the passive components which are used for decoupling filter, energy storage and management design. Such dc-dc converter modules are commercialized for quite a long time. For example, Texas Instruments

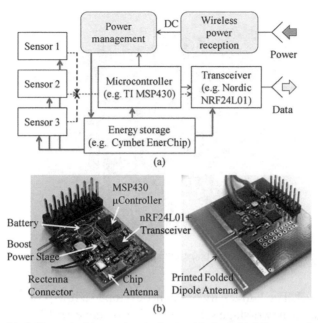

Figure 5.21 (a) Self-sustainable sensor platform with far-field RF energy harvesting unit. (b) Power management module based on commercial components [96].

has released dc-dc converter products including a low power converter [95]. This low power converter allows voltage boosting from 1.5V to 7.5V. Self-sustainable sensor platform with far-field RF energy harvesting unit and its power management module based on commercial components are shown in Figure 5.21 [96].

5.5.2 Notable Lab Results

Japanese researchers have proposed a buck-boost dc-dc converter in discontinuous conduction mode (DCM) which has a boosting efficiency of more than 80% for the variation of loads ranging from 400 to 4000 Ω at a low input voltage of 0.8 V [97]. Integrating this boost converter with a class-F rectifier has an overall efficiency of 60% with loads changing from 100 to 5000 Ω.

A similar improvement has been made by Shanghai Jiaotong University on cascaded boost-buck converter as shown in Figure 5.22 [98]. With a conversion efficiency of about 90%, such a converter is suitable for an intermediate or large input power range as switching loss will be dominant in a

Figure 5.22   Cascaded boost-buck dc-dc converter reported in [98].

low power level. The University of Pavia developed a low-power management system including a two-stage self-starting boost converter [99]. With the input power varying from 2.5 μW to 1mW, this battery-free system shows a steady charge transfer efficiency of 55%.

5.6 Conclusion

Although rectenna technology is not new and early researches have been led from 70s to 90s on rectenna principle at watt and milliwatt level, present scenarios are now pushing innovation toward μW level. We have seen in this chapter that at this power, rectifier efficiency can be improved in many new ways (increasing the collected energy, increasing the diode's nonlinearity, operating at lower temperature, building circuit with higher junction resistance diodes) in addition to classical ways (high efficient antenna, boost converter, low diode parasitics). The other research topic that has been raised? for the past years, due to the highly varying input power of real environment is the increase of dynamic range, that constitutes a promising challenge.

References

[1] T.-W. Yoo and K. Chang, "Theoretical and experimental development of 10 and 35 GHz rectennas," *IEEE Trans. Microw. Theory and Techn.*, vol. 40, pp. 1259–1266, 1992.

[2] K. M. K. Komurasaki, W. Hatakeyama, Y. Okamoto, S. Minakawa, M. Suzuki, K. Shimamura, *et al.*, "Microstrip antenna and rectifier for wireless power transfar at 94 GHz," in *Proc. Wireless Power Transf. Conf.*, 2017, pp. 1–3.

[3] H.-K. Chiou and I.-S. Chen, "High-Efficiency Dual-Band On-Chip Rectenna for 35-and 94-GHz Wireless Power Transmission in 0.13-μm CMOS Technology," *IEEE Trans. Microw. Theory and Techn.*, vol. 58, pp. 3598–3606, 2010.

[4] S. Hemour, C. H. Lorenz, and K. Wu, "Small-footprint wideband 94 GHz rectifier for swarm micro-robotics," in *IEEE MTT-S Int. Microw. Symp. Dig.*, 2015, pp. 1–4.

[5] N. Weissman, S. Jameson, and E. Socher, "W-band CMOS on-chip energy harvester and rectenna," in *IEEE MTT-S Int. Microw. Symp. Dig.*, 2014, pp. 1–3.

[6] H. Gao, M. K. Matters-Kamrnerer, P. Harpe, D. Milosevic, U. Johannsen, A. van Roermund, *et al.*, "A 71 GHz RF energy harvesting tag with 8% efficiency for wireless temperature sensors in 65nm CMOS," in *Proc. IEEE Radio Freq. Integr. Circuits Symp.*, 2013, pp. 403–406.

[7] H. Gao, M. K. Matters-Kammerer, D. Milosevic, A. van Roermund, and P. Baltus, "A 62 GHz inductor-peaked rectifier with 7% efficiency," in *Proc. IEEE Radio Freq. Integr. Circuits Symp.*, 2013, pp. 189–192.

[8] K. Hatano, N. Shinohara, T. Mitani, T. Seki, and M. Kawashima, "Development of improved 24 GHz-band class-F load rectennas," in *Proc. IEEE MTT-S Int. Microw. Workshop Ser. Innovative Wireless Power Transmiss.: Technol., Syst., Appl.*, 2012, pp. 163–166.

[9] N. Shinohara and H. Matsumoto, "Experimental study of large rectenna array for microwave energy transmission," *IEEE Trans. Microw. Theory and Techn.*, vol. 46, pp. 261–268, 1998.

[10] Y. Kobayashi, H. Seki, and M. Itoh, "Improvement of a rectifier circuit of rectenna element for the stratosphere radio relay system (in Japanese)," in *Proc. IEICE B*, 1993, pp. 2–37.

[11] T. Saka, "An experiment of a C band rectenna," in *Proc. SPS97*, 1997, pp. 251–253.

[12] Y. Fujino, M. Fujita, N. Kaya, S. Kunimi, M. Ishii, N. Ogihara, *et al.*, "A dual polarization microwave power transmission system for microwave propelled airship experiment," in *Proc. ISAP'96*, 1996, vol. 2, pp. 393–396.

[13] J. McSpadden, K. Chang, and A. Patton, "Microwave power transmission research at Texas A&M University," *Space Energy Transport.*, vol. 1, pp. 368–393, 1996.

[14] T. Shibata, Y. Aoki, M. Otsuka, T. Idogaki, and T. Hattori, "Microwave energy transmission system for microrobot," *IEICE Trans. Electron.*, vol. 80, pp. 303–308, 1997.

[15] L. W. Epp, A. R. Khan, H. K. Smith, and R. P. Smith, "A compact dual-polarized 8.51 GHz rectenna for high-voltage (50 V) actuator applications," *IEEE Trans. Microw. Theory and Techn.*, vol. 48, pp. 111–120, 2000.

[16] W. C. Brown, "An experimental low power density rectenna," in *IEEE MTT-S Int. Microw. Symp. Dig.*, 1991, pp. 197–200.

[17] J. O. McSpadden, L. Fan, and K. Chang, "A high conversion efficiency 5.8 GHz rectenna," in *IEEE MTT-S Int. Microw. Symp. Dig.*, 1997, pp. 547–550.

[18] R. J. Gutmann and J. M. Borrego, "Power combining in an array of microwave power rectifiers," *IEEE Trans. Microw. Theory and Techn.*, vol. 27, pp. 958–968, 1979.

[19] W. C. Brown, "The history of the development of the rectenna," in *Proc. SPS Microw. Syst. Workshop at JSC-NASA,* 1980, pp. 271–280.

[20] J. Koomey, S. Berard, M. Sanchez, and H. Wong, "Implications of historical trends in the electrical efficiency of computing," *IEEE Ann. History Comput.*, vol. 33, pp. 46–54, 2011.

[21] J. Koomey and S. Naffziger, "Moore's Law might be slowing down but not energy efficiency," *IEEE Spectrum.*, vol. 31, 2015.

[22] S. Hemour, Y. Zhao, C. H. P. Lorenz, D. Houssameddine, Y. Gui, C.-M. Hu, *et al.*, "Towards low-power high-efficiency RF and microwave energy harvesting," *IEEE Trans. Microw. Theory and Techn.*, vol. 62, pp. 965–976, 2014.

[23] Philips Semiconductors. "General purpose operational amplifier μA741/μA741C/SA741C," μA741 datasheet, Aug. 1994.

[24] Texas Instruments, "LMC6041 CMOS Single Micropower Operational Amplifier," LMC6041 datasheet, 2001.

[25] Texas Instruments, "LPV511 Micropower, Rail-to-Rail Input and Output Operational Amplifier," LPV511 datasheet, 2005.

[26] Maxim Intgrated, "MAX44264 nanoPower Op Amp in a Tiny 6-Bump WLP," MAX44264 datasheet, 2010.

[27] Touchstone Semiconductor, "TS1002–04 0.8V/0.6μA Rail-to-Rail Dual/Quad Op Amps," TS1002 datasheet, 2012.

[28] Texas Instruments, "LPV801 Single Channel 450nA Nanopower Operational Amplifier," LPV801 datasheet, 2016.

[29] H. Sun, Y.-X. Guo, M. He, and Z. Zhong, "A dual-band rectenna using broadband yagi antenna array for ambient RF power harvesting," *IEEE Antennas Wireless Propag. Lett.,* vol. 12, pp. 918–921, 2013.

[30] H. Sun, Y.-X. Guo, M. He, and Z. Zhong, "Design of a high-efficiency 2.45 GHz rectenna for low-input-power energy harvesting," *IEEE Antennas Wireless Propag. Lett.,* vol. 11, pp. 929–932, 2012.

[31] J. O. McSpadden, T. Yoo, and K. Chang, "Theoretical and experimental investigation of a rectenna element for microwave power transmission," *IEEE Trans. Microw. Theory and Techn.,* vol. 40, pp. 2359–2366, 1992.

[32] W. Brown and J. Triner, "Experimental thin-film, etched-circuit rectenna," in *IEEE MTT-S Int. Microw. Symp. Dig.,* 1982, pp. 185–187.

[33] W. C. Brown, "Optimization of the efficiency and other properties of the rectenna element," in *IEEE MTT-S Int. Microw. Symp. Dig.,* 1976, pp. 142–144.

[34] Y.-J. Ren and K. Chang, "5.8 GHz circularly polarized dual-diode rectenna and rectenna array for microwave power transmission," *IEEE Trans. Microw. Theory and Techn.,* vol. 54, pp. 1495–1502, 2006.

[35] S.-M. Han, J.-Y. Park, and T. Itoh, "Dual-fed circular sector antenna system for a rectenna and a RF receiver," in *Proc. 34th Eur. Microw. Conf.,* 2004, pp. 1089–1092.

[36] J. O. McSpadden, L. Fan, and K. Chang, "Design and experiments of a high-conversion-efficiency 5.8 GHz rectenna," *IEEE Trans. Microw. Theory and Techn.,* vol. 46, pp. 2053–2060, 1998.

[37] J. O. McSpadden and K. Chang, "A dual polarized circular patch rectifying antenna at 2.45 GHz for microwave power conversion and detection," in *IEEE MTT-S Int. Microw. Symp. Dig.,* 1994, pp. 1749–1752.

[38] S. Kim, R. Vyas, J. Bito, K. Niotaki, A. Collado, A. Georgiadis, *et al.,* "Ambient RF energy-harvesting technologies for self-sustainable standalone wireless sensor platforms," *Proc. IEEE,* vol. 102, pp. 1649–1666, 2014.

[39] C. H. P. Lorenz, S. Hemour, and K. Wu, "Physical mechanism and theoretical foundation of ambient RF power harvesting using zerobias diodes," *IEEE Trans. Microw. Theory and Techn.,* vol. 64, pp. 2146–2158, 2016.

[40] M. Pinuela, P. D. Mitcheson, and S. Lucyszyn, "Ambient RF energy harvesting in urban and semi-urban environments," *IEEE Trans. Microw. Theory and Techn.*, vol. 61, pp. 2715–2726, 2013.

[41] C. Song, Y. Huang, J. Zhou, J. Zhang, S. Yuan, and P. Carter, "A high-efficiency broadband rectenna for ambient wireless energy harvesting," *IEEE Trans. Antennas Propag.*, vol. 63, pp. 3486–3495, 2015.

[42] A. Mavaddat, S. H. M. Armaki, and A. R. Erfanian, "Millimeter-Wave Energy Harvesting Using Microstrip Patch Antenna Array," *IEEE Antennas Wireless Propag. Lett.*, vol. 14, pp. 515–518, 2015.

[43] A. Mavaddat, S. H. M. Armaki, and A. R. Erfanian, "Millimeter-Wave Energy Harvesting Using 4 × 4 Microstrip Patch Antenna Array," *IEEE Antennas Wireless Propag. Lett.*, vol. 14, pp. 515–518, 2015.

[44] T. S. Almoneef, H. Sun, and O. M. Ramahi, "A 3-D folded dipole antenna array for far-field electromagnetic energy transfer," *IEEE Antennas Wireless Propag. Lett.*, vol. 15, pp. 1406–1409, 2016.

[45] S. Shao, K. Gudan, and J. J. Hull, "A mechanically beam-steered phased array antenna for power-harvesting applications [Antenna Applications Corner]," *IEEE Antennas Propag. Mag.*, vol. 58, pp. 58–64, 2016.

[46] K. Noguchi, N. Nambo, H. Miyagoshi, K. Itoh, and J. Ida, "Design of high-impedance wideband folded dipole antennas for energy harvesting applications," in *Proc. IEEE 4th Asia–Pacific Conf. Antennas Propag. (APCAP)*, 2015, pp. 257–258.

[47] N. Shinohara and Y. Zhou, "Development of rectenna with high impedance and high Q antenna," in *Proc. Asia–Pacific Microw. Conf.*, 2014, pp. 600–602.

[48] M. Arrawatia, M. S. Baghini, and G. Kumar, "Broadband RF energy harvesting system covering CDMA, GSM900, GSM1800, 3G bands with inherent impedance matching," in *IEEE MTT-S Int. Microw. Symp. Dig.*, 2016, pp. 1–3.

[49] C. Song, Y. Huang, J. Zhou, and P. Carter, "Improved ultrawideband rectennas using hybrid resistance compression technique," *IEEE Trans. Antennas Propag.*, vol. 65, pp. 2057–2062, 2017.

[50] H. Saghlatoon, T. Björninen, L. Sydänheimo, M. M. Tentzeris, and L. Ukkonen, "Inkjet-printed wideband planar monopole antenna on cardboard for RF energy-harvesting applications," *IEEE Antennas Wireless Propag. Lett.*, vol. 14, pp. 325–328, 2015.

[51] S. Korhummel, D. G. Kuester, and Z. Popoviæ, "A harmonically-terminated two-gram low-power rectenna on a flexible substrate," in *Proc. Wireless Power Transf. Conf.*, 2013, pp. 119–122.

[52] Y.-S. Chen and C.-W. Chiu, "Maximum Achievable Power Conversion Efficiency Obtained Through an Optimized Rectenna Structure for RF Energy Harvesting," *IEEE Trans. Antennas Propag.*, vol. 65, pp. 2305–2317, 2017.

[53] C. Song, Y. Huang, J. Zhou, P. Carter, S. Yuan, Q. Xu, *et al.*, "Matching network elimination in broadband rectennas for high-efficiency wireless power transfer and energy harvesting," *IEEE Trans. Ind. Electron.*, vol. 64, pp. 3950–3961, 2017.

[54] E. Vandelle, P. Doan, D. Bui, T. Vuong, G. Ardila, K. Wu, *et al.*, "High gain isotropic rectenna," in *Proc. Wireless Power Transf. Conf.*, 2017, pp. 1–4.

[55] I. J. Bahl and D. K. Trivedi, "A Designer's Guide to Mcrostrip Line," *Microwaves*, 1977.

[56] D. M. Pozar, *Microwave Engineering*. Hoboken, NJ, USA: John Wiley & Sons, 2009.

[57] A. M. Niknejad, *Electromagnetics for High-Speed Analog and Digital Communication Circuits*, 1st ed. Cambridge, U.K.: Cambridge Univ. Press, 2007.

[58] H. W. Bode, *Network analysis and feedback amplifier design*, Princeton, NJ, USA: Van Nostrand, 1945.

[59] R. M. Fano, "Theoretical limitations on the broadband matching of arbitrary impedances," *J. Franklin Inst.*, vol. 249, pp. 57–83, 1950.

[60] C. H. P. Lorenz, S. Hemour, W. Li, Y. Xie, J. Gauthier, P. Fay, et al., "Breaking the Efficiency Barrier for Ambient Microwave Power Harvesting With Heterojunction Backward Tunnel Diodes," *IEEE Trans. Microw. Theory and Techn.*, vol. 63, pp. 4544–4555, 2015.

[61] A. Collado and A. Georgiadis, "24 GHz substrate integrated waveguide (SIW) rectenna for energy harvesting and wireless power transmission," in *IEEE MTT-S Int. Microw. Symp. Dig.*, 2013, pp. 1–3.

[62] C. H. Petzl Lorenz, "Mécanismes physiques et fondements théoriques de la récupération d'énergie micro-ondes ambiante pour les dispositifs sans fil à faible consommation," M.Sc., Electr. Eng. Dept., Polytechnique Montréal, Montreal, Canada, 2015.

[63] D. Wang, M.-D. Wei, and R. Negra, "Design of a broadband microwave rectifier from 40 MHz to 4740 MHz using high impedance inductor," in *Proc. Asia–Pacific Microw. Conf.*, 2014, pp. 1010–1012.

[64] M. Arrawatia, M. S. Baghini, and G. Kumar, "Broadband Bent Triangular Omnidirectional Antenna for RF Energy Harvesting," *IEEE Antennas Wireless Propag. Lett.*, vol. 15, pp. 36–39, 2016.

[65] D. Wang, X. A. Nghiem, and R. Negra, "Design of a 57% bandwidth microwave rectifier for powering application," in *Proc. Wireless Power Transf. Conf.*, 2014, pp. 68–71.

[66] M.-J. Nie, X.-X. Yang, G.-N. Tan, and B. Han, "A compact 2.45 GHz broadband rectenna using grounded coplanar waveguide," *IEEE Antennas Wireless Propag. Lett.*, vol. 14, pp. 986–989, 2015.

[67] T. W. Barton, J. M. Gordonson, and D. J. Perreault, "Transmission line resistance compression networks and applications to wireless power transfer," *IEEE J. Emerg. Sel. Topics Power Electron.*, vol. 3, pp. 252–260, 2015.

[68] D. Masotti, A. Costanzo, P. Francia, M. Filippi, and A. Romani, "A load-modulated rectifier for RF micropower harvesting with start-up strategies," *IEEE Trans. Microw. Theory and Techn.*, vol. 62, pp. 994–1004, 2014.

[69] Z. Liu, Z. Zhong, and Y.-X. Guo, "Enhanced Dual-Band Ambient RF Energy Harvesting With Ultra-Wide Power Range," *IEEE Microw. Wireless Compon. Lett.*, vol. 25, pp. 630–632, 2015.

[70] A. Almohaimeed, M. Yagoub, and R. Amaya, "Efficient rectenna with wide dynamic input power range for 900 MHz wireless power transfer applications," in *Proc. IEEE Elect. Power Energy Conf. (EPEC), 2016 IEEE*, 2016, pp. 1–4.

[71] S. H. Abdelhalem, P. S. Gudem, and L. E. Larson, "An RF–DC converter with wide-dynamic-range input matching for power recovery applications," *IEEE Trans. Circuits Syst. II, Exp. Briefs,* vol. 60, pp. 336–340, 2013.

[72] X. Y. Zhang, Z.-X. Du, and Q. Xue, "High-Efficiency Broadband Rectifier With Wide Ranges of Input Power and Output Load Based on Branch-Line Coupler," *IEEE Trans. Circuits Syst. I, Reg. Papers,* vol. 64, pp. 731–739, 2017.

[73] H. C. Torrey and C. A. Whitmer, *Crystal Rectifiers*, New York, NY, USA: McGraw-Hill, 1948.

[74] S. Herner, A. Weerakkody, A. Belkadi, and G. Moddel, "High performance MIIM diode based on cobalt oxide/titanium oxide," *Appl. Phys. Lett.*, vol. 110, p. 223901, 2017.

[75] S. Hemour and K. Wu, "Reactive nonlinearity for low power rectification," in *Proc. IEEE int. Wireless Sym. (IWS),* 2015, pp. 1–4.

[76] V. Giordano, C. Fluhr, B. Dubois, and E. Rubiola, "Characterization of zero-bias microwave diode power detectors at cryogenic temperature," *Rev. Sci. Instrum.*, vol. 87, p. 084702, 2016.

[77] X. Gu, S. Hemour, and K. Wu, "Integrated cooperative radiofrequency (RF) and kinetic energy harvester," in *Proc. Wireless Power Transf. Conf.*, 2017, pp. 1–3.

[78] C. H. Lorenz, S. Hemour, W. Liu, A. Badel, F. Formosa, and K. Wu, "Hybrid power harvesting for increased power conversion efficiency," *IEEE Microw. Wireless Compon. Lett.*, vol. 25, pp. 687–689, 2015.

[79] F. Giuppi, K. Niotaki, A. Collado, and A. Georgiadis, "Challenges in energy harvesting techniques for autonomous self-powered wireless sensors," in *Proc. 43th Eur. Microw. Conf. (EuMC)*, 2013, pp. 854–857.

[80] J. Bito, R. Bahr, J. G. Hester, S. A. Nauroze, A. Georgiadis, and M. M. Tentzeris, "A Novel Solar and Electromagnetic Energy Harvesting System With a 3-D Printed Package for Energy Efficient Internet-of-Things Wireless Sensors," *IEEE Trans. Microw. Theory and Techn.*, vol. 65, pp. 1831–1842, 2017.

[81] A. Collado and A. Georgiadis, "Optimal waveforms for efficient wireless power transmission," *IEEE Microw. Wireless Compon. Lett.*, vol. 24, pp. 354–356, 2014.

[82] A. J. S. Boaventura, A. Collado, A. Georgiadis, and N. B. Carvalho, "Spatial power combining of multi-sine signals for wireless power transmission applications," *IEEE Trans. Microw. Theory and Techn.*, vol. 62, pp. 1022–1030, 2014.

[83] F. Bolos, J. Blanco, A. Collado, and A. Georgiadis, "RF Energy Harvesting From Multi-Tone and Digitally Modulated Signals," *IEEE Trans. Microw. Theory and Techn.*, vol. 64, pp. 1918–1927, 2016.

[84] L. Rizo, M. Ruiz, and J. García, "Device characterization and modeling for the design of UHF Class-E inverters and synchronous rectifiers," in *Proc. IEEE 15th Workshop Control Modeling Power Electron. (COMPEL)*, 2014, pp. 1–5.

[85] M. Litchfield, S. Schafer, T. Reveyrand, and Z. Popoviæ, "High-efficiency X-band MMIC GaN power amplifiers operating as rectifiers," in *IEEE MTT-S Int. Microw. Symp. Dig.*, 2014, pp. 1–4.

[86] M. Ruiz and J. García, "An E-pHEMT self-biased and self-synchronous class E rectifier," in *IEEE MTT-S Int. Microw. Symp. Dig.*, 2014, pp. 1–4.

[87] J.-W. Yang and H.-L. Do, "High-efficiency zvs ac-dc led driver using a self-driven synchronous rectifier," *IEEE Trans. Circuits Syst. I, Reg. Papers,* vol. 61, pp. 2505–2512, 2014.

[88] S. Dehghani and T. Johnson, "A 2.4-GHz CMOS Class-E Synchronous Rectifier," *IEEE Trans. Microw. Theory and Techn.*, vol. 64, pp. 1655–1666, 2016.

[89] M. Stoopman, S. Keyrouz, H. J. Visser, K. Philips, and W. A. Serdijn, "Co-design of a CMOS rectifier and small loop antenna for highly sensitive RF energy harvesters," *IEEE J. Solid-State Circuits,* vol. 49, pp. 622–634, 2014.

[90] M. Roberg, T. Reveyrand, I. Ramos, E. A. Falkenstein, and Z. Popovic, "High-efficiency harmonically terminated diode and transistor rectifiers," *IEEE Trans. Microw. Theory and Techn.,* vol. 60, pp. 4043–4052, 2012.

[91] K. Hamano, R. Tanaka, S. Yoshida, A. Miyachi, K. Nishikawa, and S. Kawasaki, "Design of dual-band rectifier using microstrip spurline notch filter," in *Proc. IEEE Int. Symp. Radio-Freq. Integr. Technol.,* 2016, pp. 1–3.

[92] M. Roberg, E. Falkenstein, and Z. Popoviæ, "High-efficiency harmonically-terminated rectifier for wireless powering applications," in *IEEE MTT-S Int. Microw. Symp. Dig.,* 2012, pp. 1–3.

[93] D. Allane, G. A. Vera, Y. Duroc, R. Touhami, and S. Tedjini, "Harmonic Power Harvesting System for Passive RFID Sensor Tags," *IEEE Trans. Microw. Theory and Techn.,* vol. 64, pp. 2347–2356, 2016.

[94] S. Ladan and K. Wu, "35 GHz harmonic harvesting rectifier for wireless power transmission," in *IEEE MTT-S Int. Microw. Symp. Dig.,* 2014, pp. 1–4.

[95] Texas Instruments, "AN-288 System-Oriented DC-DC Conversion Techniques," AN-288 datasheet, 2013.

[96] Z. Popović, E. A. Falkenstein, D. Costinett, and R. Zane, "Low-power far-field wireless powering for wireless sensors," *Proc. IEEE,* vol. 101, pp. 1397–1409, 2013.

[97] Y. Huang, N. Shinohara, and T. Mitani, "A constant efficiency of rectifying circuit in an extremely wide load range," *IEEE Trans. Microw. Theory and Techn.,* vol. 62, pp. 986–993, 2014.

[98] M. Fu, C. Ma, and X. Zhu, "A cascaded boost–buck converter for high-efficiency wireless power transfer systems," *IEEE Trans. Ind. Informat.,* vol. 10, pp. 1972–1980, 2014.

[99] E. Dallago, A. L. Barnabei, A. Liberale, G. Torelli, and G. Venchi, "A 300-mV Low-Power Management System for Energy Harvesting Applications," *IEEE Trans. Power Electron.,* vol. 31, pp. 2273–2281, 2016.

PART II

Applications

6

Far Field Energy Harvesting
and Backscatter Communication

**Saman Naderiparizi[1], Aaron N. Parks[1], Zerina Kapetanovic[1]
and Joshua R. Smith[2]**

[1]Department of Electrical Engineering, University of Washington, USA
[2]Paul G Allen School of Computer Science and Engineering, and Department
of Electrical Engineering, University of Washington, USA

6.1 Introduction

The energy efficiency of computing has been improving exponentially for
the last 70 years. In fact, today's microelectronics are about one trillion
times more energy efficient than early computers such as the Eniac [1, 2].
Around the early 2000s, this trend in energy efficiency made it possible to
power general purpose microcontrollers and low-power sensors using only
propagating radio waves [3–5]. Compared to other ways of delivering or
collecting power, RF harvesting has the benefit of only requiring an antenna
for transduction, which connected devices generally already contain; other
forms of energy harvesting require non-standard materials and components,
such as photovoltaic cells, thermoelectric junctions, or mechanical generators, not just ordinary conductors and semiconductors. Compared to solar,
thermal or vibration harvesting, ambient RF energy harvesting has an added
advantage in that it typically does not have any downtime, which in turn can
allow low-power ambient RF harvesting devices to be made battery-free.

One of the most commonly observed uses of RF energy harvesting is in
RFID systems. RFID tags, generally used in inventory management systems,
consist of a low-cost printed antenna and an RF-powered communication IC.
Readers, or interrogators, power up the tags and communicate with them,
usually transferring simple information such as the ID number of the tag.
RFID has not found widespread use beyond what essentially constitutes

143

	Planted	Wild
Near-field	NFC WISP [7] WREL [8]	Capturing 60 Hz Leakage
Far-field (this chapter)	WISP [4, 5] WISPCam [10–12]	Ambient RF Harvesting [9, 29] Ambient Backscatter

Figure 6.1 RF-powered systems can be divided into four categories: "Planted" means energy is intentionally delivered. "Wild" means the energy exists in ambient form with no need for additional infrastructure. In this work, we focus on far-field systems which harvest both Planted and Wild sources of energy. Near-field ambient energy would include leakage from 60 Hz power lines, an area that we have not thoroughly explored.

barcode replacement, but early adopters of enhanced RFID have used this technology in applications such as tire pressure monitoring for aircraft [6]. As we describe in this chapter, the RFID paradigm is a technological foundation that is ripe for pushing into domains ranging from structural health monitoring to pick-to-light systems [13] to battery-less security cameras.

In recent years, we have explored the use of RF energy delivery/ harvesting across a variety of platforms, extending further than common RFID tags. As illustrated in Figure 6.1, the source of RF energy for these platforms can be either "planted," such as an RFID reader or Near-Field Communication reader (a near-field equivalent of UHF RFID), or it can be "wild," such as energy from television broadcasting and cellular towers.

In this section, the focus is on methods and applications developed in recent years based on far-field wireless power transfer. The first system we explore is the Wireless Identification and Sensing Platform (WISP), along with one of its latest incarnations, a battery-free camera called WISPCam. WISP and WISPCam harvest energy from an RFID reader, a planted source of wireless power. We then go on to discuss ambient RF harvesting systems and Ambient Backscatter systems. These systems find creative ways to gather and use energy from television broadcasting towers, which are examples of wild wireless power sources.

6.2 Planted RF Harvesting

Recent research has shown that battery-free sensing is feasible with passive RFID technology. The main building block that enabled a lot of these sensing applications is Wireless Identification and Sensing Platform (WISP). In the following section the history behind the development of WISP will be introduced. Then one significant application, which uses the WISP to enable battery-free image capturing, will be discussed.

6.2.1 WISP

The idea of WISP started in 2004 when researchers were interested in applications targeted towards elderly care. More specifically, they wanted to find the answer to one key question: how can we build battery-free sensors that can detect when a user is interacting with an object?

α-WISP was introduced to solve this problem [3]. α-WISP used two anti-parallel mercury switches to multiplex two commercial UHF RFID tags, with two different Electronic Product Code (EPC) ID's. Basically, α-WISP forms an accelerometer with 1-bit resolution. When it is tilted in one direction it sends one ID and when tilted the other way it sends another ID to the RFID reader. α-WISP was the first RFID tag that transmitted sensor data, a use-case that RFID has not been designed for.

Later in 2005, π-WISP was introduced which was capable of sending any arbitrary sensor data [14]. π-WISP is similar to α-WISP by having two RFID chips multiplexed to an antenna with an electronic switch. The switch was controlled by a low power micrcontroller which was programmed to read sensor data. A power harvester system consisting of a diode-based Dickson charge pump was designed to provide power for the sensor and the microcontroller. The microcontroller encoded the sensor data bit-stream by alternating between two RFID chips. Since the IDs for π-WISP were produced by software (rather than the mechanical mercury switch in α-WISP), it was capable of generating any complex arbitrary bit-stream. However, the main problem was the low bit-rate of π-WISP due to the large communication overhead imposed by the long ID (usually 96 bits).

In 2006, the first version of WISP was designed and successfully implemented the full RFID Class 1 Generation 1 (C1G1) in software [5]. WISP is a programmable, RF-powered, RFID tag which uses a microcontroller to handle the protocol stack as well as interfacing with sensors. Unlike α and π-WISP that had a large communication overhead, WISP was able to use some (or all) of the tag ID's for sending sensor data. While WISP was getting more mature, RFID protocols were improving too. After introducing the EPC Class 1 Generation 2 RFID protocol, which was increasing the read rate of RFID tags, a new version of WISP firmware was developed that implemented the MAC layer of Generation 2 RFID protocol [15].

Today, the latest version of WISP is called WISP 5.1, which first off complies with the highest tag read-rate supported by commercial RFID readers, and second has a better firmware interface. In addition to this, the operation range of the WISP 5.1 improved to about 10 meters from 4 meters in its earlier versions, by using a more efficient power harvester. WISP 5.0 and 5.1 were designed at the University of Washington and were supported by the National Science Foundation.

WISP 5 is an open source platform with both its hardware design files and firmware available [16]. So far, over 300 WISP 5s have been distributed among more than 100 research institution around the world. There are numerous research prototypes available that leverage WISP 5 platform. WISPCam, which is a battery-free RFID camera, is an important research prototype that was developed on top of WISP 5 platform. Two different prototype revisions of WISPCam are shown in Figure 6.2. Next we explain the main difference between WISPCam and other battery-free sensors.

Most battery-free sensing systems target simple applications, such as sensing temperature or light intensity [29], more popular phenomena like neural signals [17], and sensing specific motion [18]. Recently, sensing systems that are based on RFID have become commercially available, spanning across a wide range of applications including but not limited to strain monitoring, pick-to-light systems, and tire pressure monitoring for aircraft [6, 13]. In recent work, researchers developed and presented the WISPCam [11, 12, 19, 20], which is believed to be the world's first completely wireless, battery-free, and RFID-based camera. In addition to RFID, it has been shown that it is possible to power sensors, including cameras, using Wi-Fi signals [21].

Camera sensors provide rich output in terms of information content comparing to other sensors. Images that cameras capture can be used in a variety of applications from surveillance to self-driving cars. Most battery-free sesnsing platforms target low-power and low data rate sensors, whereas cameras are on the opposite side of the spectrum, requiring significantly more

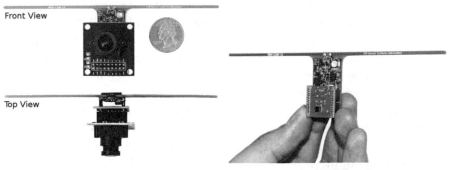

(a) Initial WISPCam Prototype   (b) Latest WISPCam Prototype

Figure 6.2   Two different WISPCam prototypes [11, 22].

power to operate and higher wireless medium data rates to transmit their information. For instance, a temperature sensor burns a few µJ of energy to provide a few bytes of data, but a camera consumes tens of milli-Joules of energy to capture an image worth of kilobytes of data [10].

In the past few years, WISPCam capabilities have expanded beyond image-capture only devices to platforms that are able to perform more complex tasks, thus are feasible for smarter scenarios. Battery-free surveillance cameras that are triggered by motion through a passive inferared sensor, a network of battery-free cameras that can localize themselves [20], and offloading computationally demanding tasks to the RFID reader to enable face detection/recognition [11] are three examples of these advancements. In the following section, WISPCam design and its applications will be discussed in more detail.

6.2.2 WISPCam

Figure 6.3 shows the high-level block diagram of the WISPCam. The harvesting system includes an antenna that absorbs RF signals produced by an RFID reader, an impedance matching network implemented using discrete components, a RF-dc conversion stage utilizing the Avago HSMS-285C Schottky barrier diodes in a single-stage RF Dickson charge pump configuration, and a high-efficiency dc–dc boost converter making use of a Texas Instruments BQ25570. A supercapacitor from the BestCap series by AVX is used to store the energy harvested from RF waves.

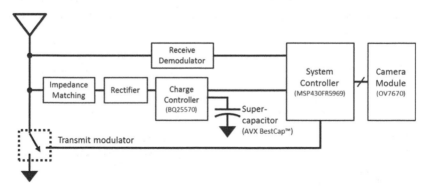

Figure 6.3 WISPCam block diagram [22]. RF wave absorbed by the antenna will pass through a power harvester consisting of an RF rectifier and a dc–dc boost converter. Harvested energy will be stored on a super capacitor. The system controller is implemented on a low-power microcontroller. The backscatter communication is based on what was implemented by WISP 5.0 [23].

Reader to tag communication (downlink) is implemented using an envelope detector and an Amplitude Shift Keying (ASK) demodulator that is power gated. The demodulator output is connected to a Texas Instruments MSP430FR5969 microcontroller that handles the EPC C1G2 protocol through running an open source firmware [23]. The microcontroller issues responses to the RFID reader queries through backscatter communication by modulating the impedance presented to the antenna using the transmit modulator switch shown.

An OVM7690 (OV7670 in the initial revision) camera by Omnivision is the sensing load of the WISPCam. Comparing with the typical byte-size quantities recorded by prior battery-free sensing loads, a camera produces relatively larger amount of data and burns more than $1000\times$ more energy per each sensing operation. The microcontroller interfaces with the camera and stores its image data onto FRAM which is a very low-power, non-volatile memory. The use of FRAM enables WISPCam with an important hibernation capability, in case the power runs out during image transmission WISPCam picks up from where it left off as soon as it harvests enough energy again without loosing any image data.

6.2.2.1 Duty-cycling

Like other battery-free platforms that are duty-cycled, WISPCam also has an intermittent operation. Basically, to power its camera module, the WISPCam's capacitor must charge to an upper threshold V_{max} through RF harvesting, then it discharges until it reaches a minimum operating voltage V_{min}. This concept is visualized in Figure 6.4(a). The amount of energy

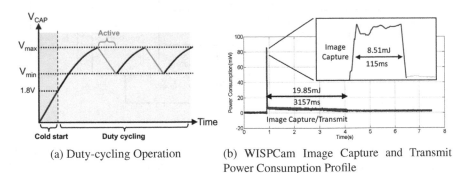

(a) Duty-cycling Operation

(b) WISPCam Image Capture and Transmit Power Consumption Profile

Figure 6.4 WISPCam operates in a duty-cycled fashion. The charge reservoir charges up to a maximum voltage V_{max} through RF energy harvesting. After reaching the V_{max} threshold, the charge reservoir discharges to a minimum voltage V_{min} [22].

delivered by the capacitor to the system during the discharge cycle must match the energy requirement of the load, in this case, the energy required to capture and transmit a photo which is shown in Figrue 6.4(b). Based on empirical data presented in Figure 6.4(b), at least 10 mJ of energy is required to capture and store a 176×144 pixel (QCIF) grayscale image (images with higher resolution will require expending more energy). Because image capture is an atomic operation, it is crucial that supercapacitor delivers at least 10 mJ to guarantee proper operation of WISPCam.

WISPCam duty-cycling frequency decreases with resepct to the input power to its harvester, which is equivalent of placing the WISPCam at a farther distance from an RFID reader. In other words, a user experiences a shorter inter-frame time before getting a new image from WISPCam if it becomes closer to the RFID reader as shown in Figure 6.5.

6.2.3 Applications

WISPCam enables battery-free imaging via harvesting energy from RF waves and transmits images using low-power backscatter communication. WISPCam also can benefit from its low-power microcontroller to process the images it captures, but its computational capability is limited due to power constraints.

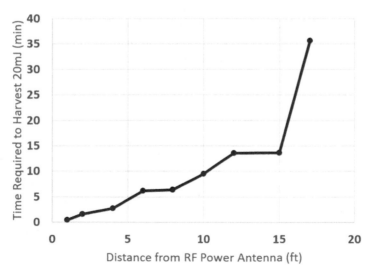

Figure 6.5 WISPCam inter-frame time versus distance from the RFID reader [22].

To support a wide range of applications, WISPCam requires some intelligence to partition its operation efficiently and spread its workload to the on-board computer and host PC (cloud). In contrast, if a device like the WISPCam lacks the intelligence to capture images based on a target application, it can potentially keep collecting and communicating data with insignificant information. This not only reduces the network-wide performance, but also increases the average power consumption of the WISPCam. For instance, in a scenario where a WISPCam is used as a surveillance camera, it should only capture an image whenever some form of movement is occurring in its field of view.

Figure 6.6 shows the high-level block diagram of WISPCam system deployment and how we can balance tasks between WISPCam and cloud allowing WISPCam to overcome its computational limitations. In essence, WISPCam can compress its backscatter data by either light computations or application based low-power triggering inputs. Then in a higher level process, heavy computational tasks can be offloaded to the cloud to overcome computational and data storage limitations.

In the following, we divide the applications that are enabled using WISPCam into two main categories: a) computationally light and b) computationally demanding applications.

6.2.3.1 Computationally light applications
Taking advantage of a power constrained on-board microcontroller, WISPCam can perform simple tasks that require no or minimal processing on the captured image prior to transmit. Some of these applications are discussed below.

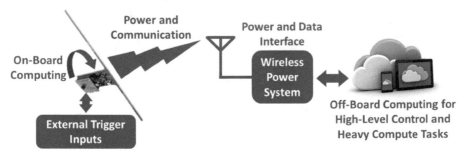

Figure 6.6   High-level block diagram of WISPCam deployment with its necessary components allowing to offload computationally demanding tasks into the host PC (cloud) [11].

6.2.3.1.1 *Analog gauge monitoring*

In the microelectronics industry, there are semiconductor fabrication plants that utilize numerous analog gauges. Despite the fact that these analog gauges are often located in hard-to-access places, they require frequent monitoring. The WISPCam can be an effective solution for monitoring these gauges because of its battery-free and wireless nature, resulting in a very low-maintenance and cost-efficient system. Similarly, monitoring analog gauges on gas tanks can be useful—for example, surveying an analog pressure gauge on a helium tank to determine when a replacement is needed. A setup in which WISPCam faces toward a helium tank analog gauge as well as some sample images captured by WISPCam are shown in Figure 6.7.

(a) Deploying a WISPCam Facing Toward an Analog Gauge

(b) Sample Image 1   (c) Sample Image 2   (d) Sample Image 3

Figure 6.7 Using WISPCam to survey analog gauges [22].

6.2.3.1.2 *Surveillance camera*

WISPCam can also be used for security and surveillance. The WISPCam is advantageous compared to current surveillance systems because it is wireless and battery-free thus it eliminates the potential undesirable scenarios such as power disconnection that would disable the system. Additionally, a Passive Infrared (PIR) motion sensor is integrated with the WISPCam to trigger image capture only when movement is detected. The PIR sensor burns only a few micro-watts of power, thus it can be always powered from the power harvester. Whenever the PIR detects motion in its environment, it triggers the image capture and transmit pipeline on WISPCam. This capability is useful in scenarios that require monitoring unexpected changes in the environment. To validate the applicability of WISPCam in such scenarios, it was faced toward a half-way open door with the PIR sensor always active. Figure 6.8 shows a sample captured image by the WISPCam that was triggered by the PIR sensor as well as the ground-truth image captured by a webcam.

6.2.3.1.3 *Self-localizing cameras*

Localization is a key problem that allows us to take more advantage of a sensor node. This problem is more challenging when targeting battery-free tags due to power and computational constraints. Some applications that are location dependent can be enabled by knowing the location where the image is captured from. A few examples include: environmental modeling,

(a) Ground-truth image from the personal camera   (b) Image captured by surveillance WISP-Cam

Figure 6.8   (a) is showing the ground-truth image captured with a web-cam camera, and (b) is demonstrating the image captured by WISPCam which was triggered by a PIR sensor [11].

3D object reconstruction, and location triggered image capture that can be enabled with a location-aware WISPCam.

The on-board camera of WISPCam can be reused to optically localize itself by having an optical clue in the field of view. The projection of n points in the image plane, whose three dimensional coordinates are known in the real world, can be used as the input to the Perspective-n-Point (PnP) algorithm which then provides the pose estimation of the camera. Generally, for $n \geq 3$ the PnP problem can be solved, but the particular case of $n = 3$ introduces some ambiguity as it may not return a unique solution. As a result, $n = 4$ is picked for localizing WISPCam. To enable this, four LEDs that are powered and controlled by a battery-free RFID tag (LED-WISP) were placed in four known locations in the field of view of the WISPCam. To detect the coordinates of LEDs in the image plane, the WISPCam captures two back-to-back images. The first image is called foreground, that has the four LEDs on, and the second image is called background, that has the four LEDs off. WISPCam can then simply subtract these two images and search for the four bright spots in the resulting image which correspond to the LEDs coordinates. Instances of $foreground$, $background$, and $foreground - background$ images are shown in Figure 6.9.

The LED-WISP continuously snifs the communication between the WISPCam and the RFID reader. Once the LED-WISP detects transmission of a secure key, known to both the WISPCam and LED-WISP, it resets its timer and synchronizes with the WISPCam. At this point, the LED-WISP knows exactly when the WISPCam is going to capture $foreground$ and $background$ images, so it can guarantees that the four LEDs are active when the WISPCam is taking the $foreground$ image and the LEDs are disabled when the WISPCam is capturing the $background$ image. A sample setup for WISPCam, LED-WISP, LEDs, and RFID reader is shown in Figure 6.9(a).

Leveraging the on-board computational capabilities available on WISP-Cam reduces total amount of data required to be sent to the RFID reader from two 140×144 images (worth of about 40KB of data) to only four coordinates (12 bytes worth of data). This provides more than $3300\times$ reduction in bandwidth requirement, which is an important factor when targeting a dense network of self-localizing battery-free cameras.

To demonstrate applicability/feasibility of this localization method in a case that contains a network of WISPCams, three WISPCams were placed in arbitrary locations such that the LED-WISP falls in their field of view. Figure 6.10 shows our experimental setup, the cameras (marked 1–3) and

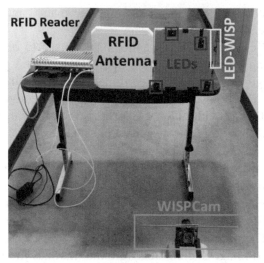

(a) Localization setup

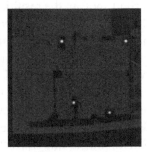

(b) Foreground Image   (c) Background Image   (d) Subtracted Image

Figure 6.9   Localization setup and a foreground, background and subtracted image samples of WISPCam. Instead of sending the entire image, WISPCam transmits only four LED coordinates which are detected through subtracting background image from the foreground [11, 20].

their views. All of the WISPCams and LED-WISPs were powered wirelessly using an RFID reader. After each WISPCam calculates the four LED's coordinates in its image plane, it sends the coordinates along with a unique ID to the RFID reader. The unique ID allows the RFID reader to uniquely identify each WISPCam.

Using this setup, WISPCams can be localized either in run-time or during setup. After the localization is done, the location information can be transmitted to WISPCams so that they are aware of their pose and location.

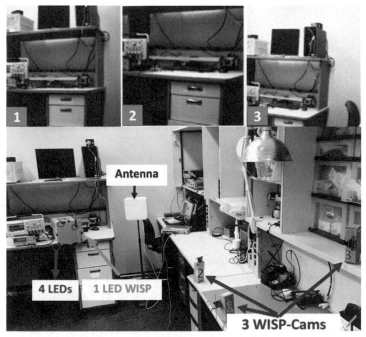

Figure 6.10 A network of three WISPCams [20].

This allows intelligently commanding any WISPCam in the network to capture a high resolution image of a particular region or zone in the environment based on a criterion such as direction of view.

6.2.3.2 Computationally demanding applications

Computer vision algorithms are known to be computationally intensive. As an example, the Viola-Jones face detection algorithm takes approximately 426 ms on a 2.2 GHz notebook to detect faces for a VGA image [24]. So if the Viola-Jones algorithm is implemented on an energy and compute constraint platform like WISPCam, a poor performance is expected. Some other computer vision tasks are even more computationally demanding such as face recognition, so enabling them by WISPCam-like platforms requires a collaboration with the host PC or cloud-based computer. Here we discuss how face detection and recognition can be done using WISPCam.

It is shown that face recognition generally has a higher recall rate on images of greater resolution up to a resolution threshold [25]. However, the total amount of data memory available on the WISPCam is just enough for an

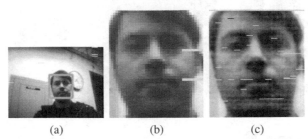

(a)   (b)   (c)

Figure 6.11   Host PC can detect the face from a low resolution 160×120 image that WISPCam captures which is shown in Firgure 6.11(a). Zooming in to extract the face from this low resolution image results in a poor quality image shown in Figure 6.11(b). To solve this problem, WISPCam captures a windowed high resolution image from the face using the face's coordinates it received from the host PC [11].

image that has a resolution 10 times lower than a VGA image. So the idea of taking a high resolution image with WISPCam and then sending it to a host PC for face recognition is not feasible given memory limitations.

On the other hand, face detection can work properly on relatively lower resolution images in comparison to face recognition [26]. So to enable face recognition, WISPCam can take a sub-sampled low resolution 160×120 image and then send it to host PC. The host PC can calculate the face coordinates and send them to WISPCam. With this information, WISPCam captures a high resolution but windowed image of the face and transmits it back to the PC. The result of this process is shown in Figure 6.11. The face captured by WISPCam has four times more pixels density in each dimension than the initial 160×120 image.

6.3 Ambient RF Harvesting

The WISP and WISPCam devices described in the previous sections all make use of planted (intentionally delivered) power. In order to operate such a device, an intentional transmitter must be placed near the point of use. In the case of an RFID system such as the WISP, these transmitters are generally RFID readers, which are complex and expensive systems resembling RADAR units.

To allow operation outside the field of intentionally-placed sources, we have surveyed other sources of energy in the environment. Solar harvesting is a commonly used technique, but only works in daylight or lighted areas.

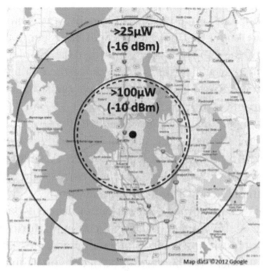

Figure 6.12 One of several 1 MW transmitters in Seattle, WA was our targeted source for early work on ambient RF harvesting. Assuming 2 dBi TX antenna gain, and 6 dBi RX antenna gain at 539 MHz,the contours in this map indicate the availability of power for RF harvesting at varying distances from the TV tower. The 10.4 km radius denoted by the dotted line is the distance at which a sensor node developed by our team was able to start up and operate [27].

Solar, piezoelectric, and thermal harvesting techniques all require expensive transducer elements. Ambient radio signals, while small in magnitude (Figure 6.12 shows how much power can be harvested at varying distances from a specific TV broadcast transmitter), are a promising source of power because they are pervasive in the built environment and because they are generally present at all hours of the day. Previous work showed us that ambient RF signals were capable of operating a system such as a sensor node [28, 29].

If a practical ambient RF-powered device could be achieved, many of the benefits in simplicity and economies of scale of standard RFID tags could be captured [4, 30]. However, unlike RFID, this device would be able to operate in any location with sufficient ambient power, reducing the need for special infrastructure and increasing the application space enormously over RFID.

6.3.1 Building a Power Supply

To begin addressing the challenge of harvesting ambient power, we first looked at the basic requirements of such a power supply.

Because the ambient signal power is typically very small compared with the load power, we know that our system will be asleep and "charging" for a large majority of the time. Thus we find that the problem of building an ambient RF-operated power supply is in essence the challenge of collecting sufficient charge to perform useful work on a charge storage medium such as a battery, supercapacitor, or capacitor.

Batteries and supercapacitors are effective at storing large amounts of energy, but unfortunately are leaky and thus can easily consume more power than is available from an ambient RF harvester. A small general purpose ceramic capacitor, with low capacitance and correspondingly low leakage, is a much more suitable choice for ambient harvesting.

We choose this small storage element out of necessity, and that choice in turn constrains the energy budget of the application platform (such as a sensor node). As an aside, much of the most interesting and most challenging work of developing an ambient RF harvesting device involves finding ways to get a task done with less energy, such that the device does not deplete the charge on the small power supply storage capacitor. We will revisit this later in the section when discussing Ambient Backscatter.

6.3.1.1 Pulling power from the air

WiFi transceivers, cellular base stations, AM/FM radio transmitters, TV broadcast transmitters, and more can all be considered potential targets for ambient RF harvesting.

We have primarily focused on TV signals due to their higher power availability and wide area coverage, but some early experimentation has been done with other signals such as cellular base transceiver station (BTS) transmissions.

In order to capture power from an ambient signal, an antenna is connected to a rectifier consisting of an arrangement of RF diodes. Because the signal strength is generally very low (below −10 dBm, or 100 μW), the voltage produced by a simple half-wave or full-wave diode rectifier would not be sufficient to operate modern logic devices, which usually require between 1v and 3.3v to operate. There are at least two differing strategies that we identify in [27] for achieving sufficient voltage:

1. **RF charge pumping** A Schottky diode-based rectifier incorporating multiple stages of voltage multiplication can produce 2–3v at sufficient input power.
2. **Post-rectification boosting** A simple full wave rectifier produces a low voltage, which is then boosted by some type of dc–dc conversion step.

In the Type 1 harvester described above, the rectifier topology generally used is the Modified RF Dickson topology illustrated in Figure 6.13. The RF Dickson rectifier/charge pump uses the oscillatory nature of the RF input signal itself to pump charge throughout a multi-stage voltage multiplying structure.

This structure, while simple to construct, may not be optimal. The reverse conduction time of diodes becomes a more significant factor in leakage as charge pump operating frequency increases, and so this structure which relies on many diodes operating at UHF frequencies can suffer from severe leakage. Additonally, if large voltage gains are required and thus many diodes are used, the diode drops can become overwhelming. At some point, adding stages to this multi-stage charge pump ceases to increase the achievable open-circuit output voltage and begins decreasing it, while also greatly impacting efficiency in a matched-load condition.

The Type 2 harvester first converts the RF signal to dc using a single full-wave rectifier (the same topology as one cell of the RF Dickson charge pump from the Type 1 system). Voltage gain is then achieved by a dc-to-dc conversion step. Figure 6.14 illustrates this topology. While more complex, this can cure some of the woes of the Type 1 system by allowing charge

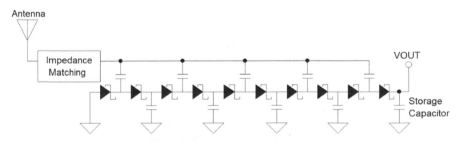

Figure 6.13   Type 1 - Five-stage RF Dickson charge pump harvester [27].

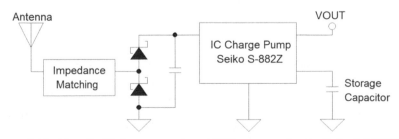

Figure 6.14   Type 2 - IC charge pump based RF harvester using the Seiko S-882Z [27].

pumps to operate at lower, more reasonable frequencies or even allowing use of low-loss inductor-based boost conversion.

Some experimental results from the Type 1 and Type 2 harvesters (Figure 6.15) show how they differ in performance. The dc–dc conversion device used in the Type 2 system, a Seiko S-882Z integrated circuit, was a very lossy and low-efficiency charge pump but did have the benefit of very good voltage gain and input sensitivity, allowing the harvesting system to start up and operate when the rectifier output was as low as 300 mV, translating in our experiments to an RF power level of around −18 dBm. In our testing, this was the best achievable sensitivity with a commercially available dc–dc converter at the time [27].

6.3.1.2 An ambient RF-powered sensor node

To demonstrate a practical application of ambient RF harvesting, we constructed an energy harvesting sensor node using both the Type 1 and Type 2

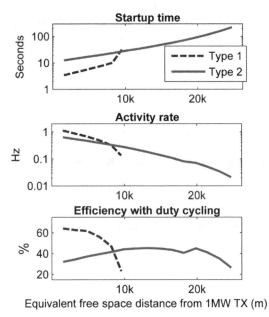

Figure 6.15   Cold start time (time required to start-up system when all nodes begin at zero Volts), activity rate, and efficiency for both harvester topologies vs. emulated distance. The Type 1 and Type 2 harvester sensitivities were −8.8 dBm and −18 dBm, respectively. Efficiency was from 23% to 64% for Type 1, and from 26% to 45% for Type 2. (Uses free-space model, 2 dBi TX gain and 6 dBi RX gain at 539 MHz, C = 160 μF ceramic chip capacitor for both harvesters) [27].

harvesters described in the previous section. Figure 6.17(a) shows the sensor platform with Type 1 harvester, plus the early prototype Type 2 harvester.

A block diagram of the node is shown in Figure 6.16. This ambient RF-scavenging node includes an onboard microcontroller, some sensing capability, and the ability to transmit sensor measurements or other data from an onboard CC2500 2.4 GHz radio module (a common and fairly energy efficient radio solution).

We programmed the onboard microcontroller to sample a light level sensor (photometer) and transmit the corresponding sensor data in the payload

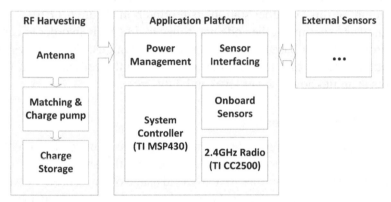

Figure 6.16   Block diagram of the ambient RF-powered sensor node [27].

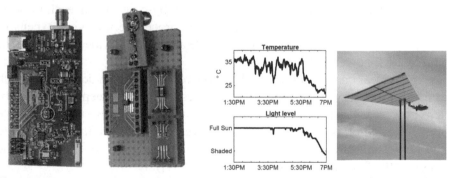

Figure 6.17   (*Left*) The ambient RF-powered wireless sensing platform, with Type 1 harvester (*far left*) and early prototype of Type 2 harvester [27]. (*Right*) Photometer and temperature measurements from the ambient RF harvesting sensor node, reported about once per second over a 5.5 hour interval. A clear cooling trend is seen in the afternoon when the node is cast in shadow [27].

of a packet using the CC2500 radio. We see that each Sense →Transmit operation requires roughly 220 µJ of energy, with most of that going to the radio transmit operation (CC2500 transmit power was +1 dBm).

The ambient sensor node was tuned for 539 MHz in order to capture power from KZJO-TV, a television station which transmits at 1 MW not far from the University of Washington campus in Seattle. The node was connected to a standard UHF TV antenna. Over a 5.5 hour length of time throughout the afternoon, the node using the Type 2 harvester was able to transmit ambient light level and temperature at a rate of nearly 1 Hz when placed at a 4.2 km distance from the KZJO-TV transmitter. Data captured and a picture of this "weather station" is reprinted in Figure 6.17(b).

In another experiment, the node was tuned to 738 MHz in order to capture power from a cellular base station (cell tower) on campus. The node was able to operate at a range of a few hundred meters from the cellular tower, which transmits at an unknown power level and duty cycle.

6.3.2 Multiband Harvesting

After proving the basic feasibility of operating sensor nodes from ambient harvesting, we refocused on addressing some of the shortcomings of the system. One of the main hurdles to real-world usage of ambient RF harvesting was the need for this system to be tuned to a particular frequency, such as that of a TV channel or cell tower transmission, in order to operate. Because the availability of certain spectral bands differ from location to location, this static tuning must occur on a case-by-case basis at each location of deployment. Tuning is a difficult manual procedure involving expensive equipment. Obviously, this is not a scalable approach to wide deployment of ambient RF-powered devices.

In an ideal case, an ultra-wideband harvester would be developed which can harvest energy at any and all frequencies. In looking deeper at this problem however, it was determined that fundamental limits exist in impedance matching, which is needed to transfer power efficiently from the antenna to the harvesting circuit (the Bode-Fano criteria imply an upper limit on the bandwidth of an effective impedance match [31]).

The solution we developed overcame this limit by rethinking the topology of the harvester. Instead of trying to build one rectifier with an ultra-wideband match to an antenna, several rectifiers were built and each was matched independently at adjacent frequency bands [9]. Figure 6.18 illustrates the basic concept of multiband harvesting.

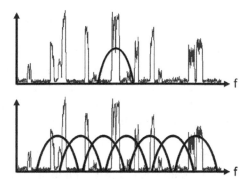

Figure 6.18 A single-band harvester cannot efficiently collect energy over a large bandwidth. The proposed multiband harvester aims to divide and conquer an arbitrarily large bandwidth for efficient harvesting [9].

The topology of the harvester is shown in Figure 6.19, where each rectifier is a Type 1 RF Dickson charge pump as shown in Figure 6.13. Each rectifier is isolated from the others by a serial bandpass filter, which is essential to the operation of the multiband device as it allows impedance matching to occur independently for each band. While we don't go through a detailed analysis of the system (which is very difficult due to non-linear behavior of the diodes) we hypothesize that N discrete harvesters can be well matched to a single

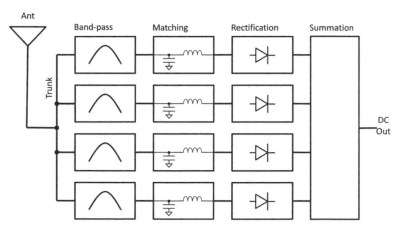

Figure 6.19 The topology of the proposed multiband harvester. A single wideband antenna is used, in contrast to other work where multiple antennas or multi-port antennas were used. Bands are spaced geometrically with a frequency ratio R [9].

antenna at N discrete frequency points, making it possible to capture energy from N frequencies at once without efficiency decrease.

In departure from prior work on multiband harvesting, our design used only one antenna and one port to capture energy from multiple bands [32, 33]. Also in departure from prior work, we decided it was of great importance that the multiband harvester be able to capture and use energy from all bands simultaneously and in such a way that they are efficiently combined.

6.3.2.1 Power combining

Combining energy from the bands is not as trivial as it may seem. Prior work had simply connected the dc output of the bands serially, a technique that worked well when all bands were excited but did not perform well when one or more bands was left unexcited (for instance, if the harvester were used in an area where one or more TV channels was not available). The issue is that, when one band is not excited, the diodes which make up its rectifier or charge pump became "dead weight", and the diode voltage drops reduced efficiency or voltage sensitivity of the system. For a many-band harvester, this efficiency and sensitivity loss would be too large for practical use.

The technique we identified to get around this was to make use of a network of "shortcut diodes," essentially diodes which allowed each unexcited stage of the harvester to be bypassed by one or fewer diode drops. The network of shortcut diodes used to boost efficiency and sensitivity when combining power from multiple bands is shown in Figure 6.20. In our testing, increases in combined output power of well over 100% were achieved using the shortcut diode network when bands were sparsely excited. Follow-up work introduced the use of switches instead of diodes, an even lower-loss solution [34].

6.3.2.1.1 *Prototyping the multiband harvester*

A 2-band and 5-band prototype are implemented with discrete components, as shown in Figure 6.21(a) and 6.21(b). The prototypes use 3 stage Dickson charge pumps in each band, use the described "shortcut" summation topology, and when the experiment requires they are connected to a wideband log-periodic antenna with a roughly 6 dBi gain. The two-band harvester design frequencies are 539 MHz and 915 MHz, and target an ambient television signal near our campus and an RFID reader, respectively. The five-band harvester design frequencies were chosen by selecting a fixed ratio of 1.5 between adjacent frequency bands, and are 267 MHz, 400 MHz, 600 MHz, 900 MHz, and 1.35 GHz.

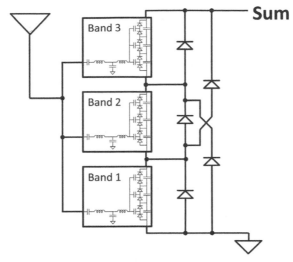

Figure 6.20 An example of the diode shortcut summation network, shown for a 3-band harvester. The output voltage will be the global sum of the band outputs, minus a small number of diode drops which is generally less than the number of unexcited bands [9].

Figure 6.21 Two multiband harvester prototypes were constructed: A 2-band harvester (a), and a 5-band harvester with diode summation network (b). The 5-band prototype measures 1.5 by 1.2 inches [9].

The 2-band and 5-band prototypes were subjected to a single-tone excitation at a power level of –10 dBm (100 µW), and the S11 (reflected power) and RF-dc conversion efficiency were measured. These single-tone results are shown in Figures 6.22(a) and 6.22(b). While these early-stage prototypes don't always perform optimally (for instance, the highest-frequency band of the 5-band harvester did not work well), we feel the results still show promise.

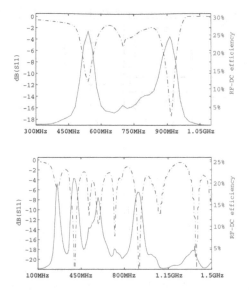

Figure 6.22   S11 (reflected power at RF port) and RF-dc conversion efficiency over the operating band for both the 2-band harvester (a) and 5-band harvester (b), for a –10 dBm test power and a 100 kΩ load [9].

6.3.3 Ambient Backscatter

After addressing the challenge of efficiently harvesting at many frequencies and thus allowing devices to work in many locales without manual retuning, we then focused on another side of the problem: The energy consumption of the sensor node. The energy harvesting sensor node which used a CC2500 radio could in fact store enough energy to communicate sensor data to the outside world, but the packets transmitted were very short and infrequent, and in practical use there might be minutes or even hours of twaiting for the charge reservoir to recharge between sensor updates. Reducing the energy requirements of the node itself could make this time shorter, enabling even more practical applications of ambient RF energy harvesting.

We note that wireless communications generally makes up the largest fraction of energy consumption in a wireless sensor node. If there were a way to reduce that fraction, it might not only be useful for ambient harvesting systems but also for sensor networks in general.

Much like harvesting, the way we addressed this was also by leveraging the ambient radio signals. In our technique, which we have titled Ambient

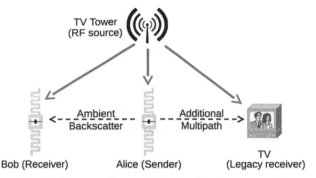

Figure 6.23 Ambient Backscatter allows communication between two battery-free devices. Devices harvest energy needed to operate from the ambient signal while modulating their reflection of that same energy in order to communicate with each other. To legacy receivers such as TV receivers, this signal is simply an additional source of multi-path, and they can still generally decode the original transmission [35].

Backscatter [35], the wireless sensor node reflects ambient radio signals to send messages, rather than generating its own signals. The RF topology for this paradigm is shown in Figure 6.23. The power required to do this modulated reflection (backscatter) operation is extremely small, three to four orders of magnitude smaller than the power required to transmit a radio signal, and thus the fraction of the sensor node's budget which must be devoted to communication now shrinks to a very small number, generally less than 1% depending on implementation.

To show the feasibility of our design, we first built simple prototypes (shown in Figure 6.24) and achieved information rates of 1 kbps over distances of up to 2.5 feet, a very modest distance [35]. Follow-on work showed that rates of 1 Mbps were achievable and that distances could be stretched to 20 m or more through the use of ultra low power multi-antenna processing and coding gain [36].

6.3.3.1 Summary

The use of ambient RF signals both as a source of power, and as a medium for near-zero power wireless communication holds promise and warrants further study. A community of researchers has now coalesced around this idea and many papers are published each year describing new insights in both ambient RF harvesting and ambient backscatter communication.

An exciting future avenue for exploration will be the combination of the ambient backscatter technique with multiband harvesting, hopefully enabling

Figure 6.24   The first Ambient Backscatter prototype device. These small card-like devices can transmit and receive data, sense, and compute, without need for a battery. The A, B, and C buttons are touch sensors, and along with two LEDS these form a simple user interface [35].

systems capable of communicating with each other at very low power in any locale.

6.4 Conclusion

The dramatic scaling of the energy efficiency of microelectronics, together with improvements in RF energy harvesting and ultra-low-power backscatter-based communication, are enabling a world in which devices can operate indefinitely without changing batteries, and without a wired connection to the power grid. We have demonstrated that even relatively high power devices such as cameras can be operated in this fashion. Our recent work has demonstrated the feasibility of RF-powered battery-free devices that sense, compute, and communicate; we expect in the coming years that such systems will become increasingly mature, practical, and widely deployed.

Acknowledgments

This work took place over several years and involved many different people. We'd like to acknowledge our co-authors and contributors, including Shyam Gollakota, Alanson Sample, Vamsi Talla, Vincent Liu, Ben Ransford, Angli Liu, Yi Zhao, James Youngquist, and all the members of the Sensor Systems laboratory at the University of Washington.

We'd also like to acknowledge our sources of funding, which include NSF grants CNS-1305072, CNS-1407583, and EEC-1028725, the Intel Science and Technology Center for Pervasive Computing, the Google PhD Fellowship program, several Google Faculty Research Awards, and a gift from Microsoft Research.

References

[1] Koomey, J. G., Berard, S., Sanchez, M., and Wong, H. (2011). Implications of historical trends in the electrical efficiency of computing. *Ann. Hist. Comput.* 33, 46–54.

[2] Smith, J. R. (ed.). (2013). "Range scaling of wirelessly powered sensor systems," in *Wirelessly Powered Sensor Networks and Computational RFID*, (New York, NY: Springer), 3–12.

[3] Philipose, M., Smith, J. R., Jiang, B., Mamishev, A., Roy, S., and Sundara-Rajan, K. (2005). Battery-free wireless identification and sensing. *IEEE Pervasive Comput.* 4, 37–45.

[4] Sample, A. P., Yeager, D. J., Powledge, P. S., Mamishev, A. V., and Smith, J. R. (2008). Design of an RFID-based battery-free programmable sensing platform. *IEEE Trans. Instrum. Meas.* 57, 2608–2615.

[5] Smith, J. R., Sample, A. P., Powledge, P. S., Roy, S., and Mamishev, A. (2006). "A wirelessly-powered platform for sensing and computation," in *Proceedings of the International Conference on Ubiquitous Computing*, (Berlin: Springer), 495–506.

[6] Tire pressure and brake temperature systems – smartstem (2014). Available at: http://www.craneae.com/Products/Sensing/SmartStem.aspx [accessed December 2014].

[7] Zhao, Y., Smith, J. R., and Sample, A. (2015). "NFC-WISPC a sensing and computationally enhanced near-field RFID platform," in *Proceedings of the 2015 IEEE International Conference on RFID (RFID)*, San Diego CA, 174–181.

[8] Sample, A. P., Meyer, D. T., and Smith, J. R. (2011). Analysis, experimental results, and range adaptation of magnetically coupled resonators for wireless power transfer. *IEEE Trans. Ind. Electron.* 58, 544–554.

[9] Parks, A. N., and Smith, J. R. (2014). "Sifting through the airwaves: efficient and scalable multi-band RF harvesting," in *Proceedings of the 2014 IEEE International Conference on RFID (IEEE RFID)*, Orlando, FL, 74–81.

[10] Naderiparizi, S., Parks, A. N., Kapetanovic, Z., Ransford, B., and Smith, J. R. (2015). "WISPCam: a battery-free rfid camera," in *Proceedings of the IEEE International Conference on RFID (RFID 2015)*, (San Diego, CA: IEEE), 166–173.

[11] Naderiparizi, S., Kapetanovic, Z., and Smith, J. R. (2016). Battery-free connected machine vision with WISPCam. *GetMobile Mob. Comput. Commun.* 20, 10–13.

[12] Naderiparizi, S., Kapetanovic, Z., and Smith, J. R. (2016). "WISPCam: An RF-powered smart camera for machine vision applications," in *Proceedings of the 4th International Workshop on Energy Harvesting and Energy-Neutral Sensing Systems*, (Stanford, CA: ACM), 19–22.

[13] Farsens battery-free sensor solutions (2014). Available at: http:// www.farsens.com/en/battery-free-sensor-solutions [accessed December 2014].

[14] Smith, J. R., Jiang, B., Roy, S., Philipose, M., Sundara-Rajan, K., and Mamishev, A. (2005). "Id modulation: embedding sensor data in an RFID timeseries," *Lecture Notes in Computer Science*, eds M. Barni, J. Herrera-Joancomart, S. Katzenbeisser, and F. Pérez-González (Berlin: Springer).

[15] Buettner, M., and Wetherall, D. (2013). "Implementing the gen 2 mac on the intel-uw WISP," in *Wirelessly Powered Sensor Networks and Computational RFID*, ed. J. R. Smith (Berlin: Springer), 143–156.

[16] WISP 5 firmware repository (2017). Available at: http://wisp5.wiki spaces.com/WISP+Home [accessed May 2017].

[17] Yeager, D. J., Holleman, J., Prasad, R., Smith, J. R., and Otis, B. P. (2009). Neuralwisp: A wirelessly powered neural interface with 1-m range. *IEEE Trans. Biomedical Circuits Syst.* 3, 379–387.

[18] Buettner, M., Prasad, R., Philipose, M., and Wetherall, D. (2009). "Recognizing daily activities with RFID-based sensors," in *Proceedings of the 11th International Conference on Ubiquitous Computing*, Orlando, FL.

[19] Naderiparizi, S., Kapetanovic, Z., and Smith, J. R. (2017). "RF-powered, backscatter-based cameras," in *Proceedings of the 2017 11th European Conference on Antennas and Propagation (EUCAP)*, Paris, 346–349.

[20] Naderiparizi, S., Zhao, Y., Youngquist, J., Sample, A. P., and Smith, J. R. (2015). "Self-localizing battery-free cameras," in *Proceedings of the 2015 ACM International Joint Conference on Pervasive and Ubiquitous Computing* (New York, NY: ACM), 445–449.

[21] Talla, V., Kellogg, B., Ransford, B., Naderiparizi, S., Gol-lakota, S., and Smith, J. R. (2015). "Powering the next billion devices with wi-fi," in *Proceedings of the 11th ACM Conference on Emerging Networking Experiments and Technologies*, New York, NY.

[22] Naderiparizi, S., Parks, A. N., Kapetanovic, Z., Ransford, B., and Smith, J. R. (2015). "WISPCam: a battery-free RFID camera," in *Proceedings of the 2015 IEEE International Conference on RFID (RFID)* (Tokyo: IEEE), 166–173.

[23] WISP 5 firmware repository (2014). Available at: http://www.github. com/wisp/ [accessed December 2014].

[24] Wang, Q., Wu, J., Long, C., and Li, B. (2012). "P-fad: Real-time face detection scheme on embedded smart cameras," in *Proceedings of the 2012 Sixth International Conference on Distributed Smart Cameras (ICDSC)*, Hong Kong, 1–6.

[25] Wang, J., Zhang, C., and Shum, H. Y. (2004). "Face image resolution versus face recognition performance based on two global methods," in *Proceedings of the Asia Conference on Computer Vision*, Jeju Island, 48–49.

[26] Marciniak, T., Chmielewska, A., Weychan, R., Parzych, M., and Dabrowski, A. (2015). Influence of low resolution of images on reliability of face detection and recognition. *Multimedia Tools Appl.* 74, 4329–4349.

[27] Parks, A. N., Sample, A. P., Zhao, Y., and Smith, J. R. (2013). "A wireless sensing platform utilizing ambient RF energy," in *Proceedings of the 2013 Radio and Wireless Symposium (RWS)*, (Austin, TX: IEEE), 331–333.

[28] Nishimoto, H., Kawahara, Y., and Asami, T. (2010). "Prototype implementation of ambient RF energy harvesting wireless sensor networks," in *Proceedings of the Sensors, 2010 Conference,* (Kona, HI: IEEE), 1282–1287.

[29] Sample, A., and Smith, J. R. (2009). "Experimental results with two wireless power transfer systems," in *Proceedings of the 4th International Conference on Radio and Wireless Symposium Radio and Wireless Symposium*, (San Diego, CA: IEEE), 16–18.

[30] Le, T., Mayaram, K., and Fiez, T. (2008). Efficient far-field radio frequency energy harvesting for passively powered sensor networks. *IEEE J. Solid State Circuits* 43, 1287–1302.

[31] Fano, R. M. (1948). *Theoretical Limitations on the Broadband Matching of Arbitrary Impedances.* Available at: http://dspace.mit. edu/handle/1721.1/12909

[32] Keyrouz, S., Visser, H. J., and Tijhuis, A. G. (2013). "Multi-band simultaneous radio frequency energy harvesting," in *Proceedings of the*

2013 7th European Conference on Antennas and Propagation (EuCAP), (Gothenburg: EuCAP), 3058–3061.

[33] Niotaki, K., Kim, S., Jeong, S., Collado, A., Georgiadis, A., and Tentzeris, M. M. (2013). A compact dual-band rectenna using slot-loaded dual band folded dipole antenna. IEEE *Antennas Wirel. Propag. Lett.* 12, 1634–1637.

[34] Parks, A. N., and Smith, J. R. (2015). "Active power summation for efficient multiband RF energy harvesting," in *Proceedings of the 2015 IEEE MTT-S International Microwave Symposium*, (Phoenix, AZ: IEEE), 1–4.

[35] Liu, V., Parks, A., Talla, V., Gollakota, S., Wetherall, D., and Smith, J. R. (2013). "Ambient backscatter: wireless communication out of thin air," in *Proceedings of the Conference on ACM SIGCOMM 2013 SIGCOMM* (New York, NY: ACM), 39–50.

[36] Parks, A. N., Liu, A., Gollakota, S., and Smith, J. R. (2014). Turbocharging ambient backscatter communication. *SIGCOMM Comput. Commun. Rev.* 44, 619–630.

7

WPT-Enabling Distributed Sensing

Luca Roselli, Paolo Mezzanotte, Valentina Palazzi, Stefania Bonafoni, Giulia Orecchini and Federico Alimenti

Department of Engineering, University of Perugia, Italy

Abstract

This chapter is devoted to show how wireless power transfer, a technology that, after almost one century of sleeping, is living its second life, can enable the development of information and communication. In particular, this chapter inserts its enabling capability in the context of recent Internet of things and Internet of space eco-systems, and focuses, in both cases, on the benefits produced on sensing applications with specific attention to the evolutionary vision represented by distributed and pervasive sensing.

7.1 Introduction

The increasing advancement of the IoT eco-system, testified by the exponentially growing number of connected objects [1], is driving and steering the development of several disruptive technologies. So far, the biggest impact has been experienced in the field of interconnection and data management, i. e., at software or high level, but the trend towards increasingly granular diffusion of "smart objects" and relevant sensing capabilities is driving a revolution even at hardware level. The paradigm shift of hardware, in turn is posing challenges on miniaturization, cost reduction, power consumption, and integrability with objects of daily use. Accordingly, several technologies are expected to concurrently foster and support this trend. Among them, WPT appears as one of the most promising approaches to supply power to miniaturized sensor tags and sensor nodes as an alternative to impracticable grid connection and life batteries. In particular, WPT is considered an enabling technology in view of massive deployment of disposable "smart sensing pebbles", to empower large surfaces of sensor matrices as well as for the energizing of sensors not precisely located in a limited region (rooms, offices, labs, but also in harsh areas for rescue or wherever distributed temporary monitoring required).

Affinities with the technological evolution driven by IoT development can be recognized also in space applications. This field, in fact, has been garrisoned so far by "traditional" technologies, purposely conceived to face tremendous requirements posed by the inherently harsh environment represented by outer space. Indeed, due to the long term duration of past missions, the difficulty of maintenance, the need to guarantee a certain quality of service and, ultimately to assure the reliability of the whole system, space-borne equipment has been traditionally manufactured by using extremely high-cost technologies. On the other hand, the recent great development of terrestrial ICTs, mostly driven by mobile applications and ultimately by IoT perspectives, increased the divide between the performance of terrestrial and satellite technologies in terms of speed, bandwidth and so forth. Aside this situation, it has to be considered the development of micro- and nano-satellite in LEOs which ended up in the "cube-satellite" paradigm. In this new scenario, characterized by less lasting, more redundant and lower "per unit" cost missions, it is possible to envisage space-borne solutions involving technologies more relaxed in terms of reliability, while exploiting the high performance of terrestrial solutions. This trend actually enabled the so-called IoS paradigm, which similarly to IoT, is characterized by a tremendous increment of object deployment (cube satellites in this case), exploiting technologies inherited by terrestrial solutions, which are "downgraded" from the point of view of reliability. In a long-term vision it can be said that IoT and IoS are bridging and closing the technological gap between terrestrial and spatial applications [2].

This chapter will focus on WPT as a means to enable sensing and massive distributed sensing according to IoT and IoS eco-system paradigm shift. To this purpose it is coherently divided in two parts.

The first part is focused on IoT, which is characterized by a relatively more mature development than IoS. The chapter firstly deals with a brief exemplifying summary of meaningful present applications, which can benefit from WPT exploitation. Then, some examples testifying the evolution of the technologies towards massive deployment and distributed sensing are given.

About IoS, according to the less mature level of development, some general considerations and some examples proving the enabling capabilities of WPT technologies in a framework of remote sensing based on distributed cube-satellite cluster-like gathered are reported.

7.2 IoT

The Internet of things is becoming one of the leading paradigms of present ICT; such a paradigm is simply based on the realization of objects able to gather information from the environment and to transfer it to Internet without any human agency. That way, truly automated and efficient productive and maintenance processes can be established. These objects, which are equipped with electronics and own both sensing and connection capabilities, are commonly called "smart objects". Several concurrent technologies can be envisaged as enabling ones for smart objects and IoT diffusion. WPT is one of the most credited ones, due to its strategic role. Indeed, on the one hand, it represents the basis of RFID technology, which is an important platform supporting communication with smart objects. On the other hand, WPT is, by itself, an essential means to directly provide power to "things" and make them autonomous [3].

In the following, some present examples testifying the growing presence of WPT into IoT applications are given.

7.2.1 IoT Enabled by WPT (Some Examples at Present)

One of the main limits to the deployment of sensors and WSNs is that they cannot be applied in hard-to-reach environments, where batteries are not easy to be replaced or have to be replaced too often. By developing RF energy harvesting, together with WPT, the battery addiction to power wireless sensors can be strongly reduced in the near future.

As an example of the present state, some non-restrictive IoT applications that will undoubtedly benefit from the use of WPT techniques to power sensors are listed hereinafter; relevant references are reported as well just to testify the emphasis about these topics:

- Monitoring of vibrations and material conditions in buildings, bridges and historical monuments [4].
- Monitoring of combustion gases and fire conditions to define alert zones [5].

- Snow level measurement to know in real time the quality of ski tracks and allow security prevention [6].
- Control of CO_2 emissions of factories, pollution emitted by cars and toxic gases generated in farms [7].
- Earthquake early detection [8].
- Chemical leakage detection [9].
- Control of temperature inside industrial and medical fridges with sensitive merchandise [10].
- Wine quality enhancing [11].
- Etc.

The state of the art on WPT, WSN and IoT proves that researchers are moving towards solutions that can limit or eventually eliminate the use of battery.

In [12], the authors highlight that, nowadays, WSNs depend on the battery duration, which is limiting the system operation and its field of applicability. As a consequence, there is a lot of interest in creating a passive sensor network scheme in the area of IoT and space-oriented WSN systems. RF energy harvesting is presented as a means to enable the control and the delivery of wireless power to RF devices. It is moreover observed that all the devices enhanced by this technology can result permanently sealed and embeddable within different structures.

Again, in [12] authors prove the feasibility of a complete passive WSN that could potentiate IoT smart objects and space craft sensing. These WSNs, in fact, can receive a continuous flow of energy that can be radiated from a base station (or any other source strong enough to support passive transponders), so that the sensing mechanisms can be continuously powered. Accordingly, they propose a combination of backscatter modulation and WPT, using the same "diode based" approach as in a traditional RF identification circuit [13]. In this way, the cost is maintained low, but the amount of gathered energy is maximized during the backscatter communication.

In [14] the authors developed sensor nodes in which the energy harvesting module is able to adapt and collect energy from solar power, ambient radio waves, and direct WPT; the wireless networking sensor nodes are conceived for environmental monitoring. Each node consists of an optimized energy harvesting module, a SoC integrated low-power Bluetooth smart transceiver, and multi-functional sensor arrays to monitor environmental parameters. The sensor arrays include pH sensor, temperature sensor, photo-detector, electromagnetic wave detector and acoustic noise detector. The SoC processes data and transmits compressed information about environmental conditions to the hub. This platform demonstrated the concepts of combining power harvesting techniques and low-power sensors for the IoT applications.

In [15] WPT is recognized to be an attractive technology to supply power to IoT devices where battery replacement is difficult or the maintenance cost is high. Polarization switching WPT antenna is proposed in the paper. The polarization switching feature makes the proposed design a good candidate for 5G-IoT sensors powered by using wireless power transfer. The 5G-IoT network can support 50,000 users per cell and a battery can last for more than ten years. The authors also underline that the key problems of the 5G-IoT network protocol are the coordination and communication between large number of IoT devices and the base station. Using the polarization switching WPT antenna technology, the cell coverage could be separated into several partitions to wake up IoT devices within one partition of the cell to avoid the collision. This proves that even if the use of WPT techniques is still under deep research and consequent development, the technology has the potential to intensely increase the spreading of IoT technologies to an enormous variety of applications.

7.2.2 IoT Future Trajectories

As exemplified in Section 7.2.1., the paradigm shift implied by the ICT evolution towards IoT is presently fostered and supported by available sensing and communication technologies. The implementation of these technologies already makes a huge class of "objects" ready to become smart. This is evident by looking at the exponential increment of the number of objects able to acquire information from the environment and to transfer it to Internet without any human agency [16]. At this level of the evolution, the main required effort is to increase the capability of managing the growing amount of data being generated by smart objects and exchanged via Internet. This issue is synthesized with the idiomatic expression: "Big Data".

With this premise, if we look at smart objects and, more precisely, at their diffusion, we notice that it is actually limited mostly by economic considerations. In order to motivate companies to add electronics to their products, in fact, it is mandatory that the cost of the technology required to make objects smart be significantly lower than the added perceived value of the resulting smart object. At present, the introduction of electronics is limited to objects that have already a fairly intrinsic high value; think for instance to industrial machines, smart cars, smart electrical appliances, smart gadgets and so forth, already present in the market. To this extent, we could say that, in spite of the increasing number of smart objects already available, we are just at the beginning of the IoT era, because objects of real day life impact are still far from being smart.

Now the question is: which technologies can enable everyday objects to become smart and to enter the IoT eco-system beyond the already available ones?

The answer can be synthetized in three technological concepts: autonomy, recyclability, and compatibility with object manufacturing processes. With respect to the present technologies able to provide smartness, the future electronics has to be increasingly compatible with the existing manufacturing processes of objects in order to keep replacing technology investments and production costs as low as possible. Moreover, these technologies have to be recyclable and ultimately compostable, in order to minimize the environmental impact of massive implementation of IoT. Finally, next-generation sensors have to be autonomous, in order to prevent either that they have to be connected to the grid (wired sensors) or that their working life is limited by the life of batteries, as clearly stated in the paragraph above. The latter issue clearly explains the importance of WPT to potentially provide autonomy to smart objects and the trend of increasing interest in this field. It is a matter of fact that RFID and RFID sensing are of the most promising ways to provide connectivity between smart objects and the Internet. From an energetic point of view, RFID tags and RFID sensor tags [17, 18] are actually powered by using wireless transfer of power. In these cases, in fact, the signal used to interrogate the objects transfers also the energy required by the tag circuitry to operate. Recently, approaches based on a direct adoption of WPT to empower smart objects are appearing in the scientific literature [16, 19] opening the way to the evolution towards disposable smart objects and massively distributed sensing also in harsh environments.

7.2.3 Sensors for Future IoT Development, Some Examples

According to the analysis reported in the previous Section 7.2.2., a key to success for IoT is the ubiquitous and distributed deployment of autonomous, recyclable and object-compatible sensors. These requirements entail the unavoidable disposal of traditional batteries, thus imposing great challenges to the future designers. If, on the one hand, the research must focus on the efficient harnessing of alternative energy sources, on the other hand, part of the efforts must be directed to the development of ultralow-power circuits.

Recently, new sensor categories have arisen to address this issue, such as chip-less RFID [20], zero-power sensors [21] and harmonic tags [22]. Besides the specific peculiarity of each category, they are all based on the principle of the RF-carrier re-usage. A transponder is interrogated by a signal, which can be sinusoidal, narrow band or broadband on the basis of the tag architecture. The latter is modulated (linearly or by using non-linear components) by the circuitry of the on board sensor and transmitted back to a suitable receiver. That way, the wireless node does not require any signal rectification or carrier generation, with a reduction in the circuit complexity (and, ultimately, in the unit cost of the tag) and in a significant energy saving.

In such a context, however, without any biasing signals, the information encoding becomes extremely challenging, due to the forced absence of any digital circuitry, synchronization, firmware or memory storage. Requirements such as sensor accuracy and information processing must be unavoidably relaxed, expected read-range reduced, whereas all the complexity and multi-node management must be entrusted to the reader side.

That being said, these new paradigms of systems are perfectly suitable for many low demanding applications (where, for instance, the periodical verification that a certain physical parameter stays under a certain threshold is enough), do not require any maintenance and allow for the adoption of flexible, biodegradable and/or wearable materials, thus going towards the direction of both "nearly-to-invisible" and "nearly-to-zero-cost" electronics.

Here, a few examples of recently developed zero-power sensors are reported together with a detailed description of the adopted information encoding strategies.

7.2.4 Impedance-Based Sensors

In 2010, a battery-free conformal CNT-based RFID-enabled sensor node for gas sensing applications was realized [23]; the tag was designed on a flexible paper substrate. The layout is shown in Figure 7.1.

CNT composites have an electrical conductance very sensitive to small quantities of gases, and are compatible with inkjet-printing. The RFID tag was designed for the European UHF RFID band centred at 868 MHz, while the printed CNTs were SWCNT from Carbon. The impedance of the SWCNT film forms the sensor part. The antenna was printed first, followed by the layers of dispersed SWCNT as a load. Thus, in presence of a given gas, the resistance of the CNT will change causing a variation of the backscattered power from the RFID. In the experiment, when 4% consistency ammonia was

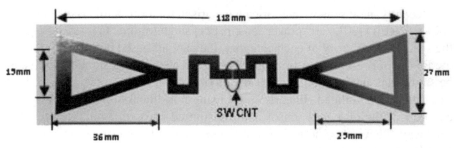

Figure 7.1   Photograph of the tag with inkjet-printed SWCNT film as a load with depicted dimensions.

imported into the gas chamber, the SWCNT impedance changed, resulting in a 10.8 dBr variation in the backscattered power from the tag antenna.

The CNT impedance was measured with a probe station to be $42.6+j11.4\Omega$ at 868 MHz. The simulation and measurement results of the return loss of the proposed antenna are shown in Figure 7.2, testifying good agreement.

The tag reveals a bandwidth between 810 MHz and 890 MHz, an omnidirectional radiation pattern with directivity around 2.0 dBi and 94.2% radiation efficiency.

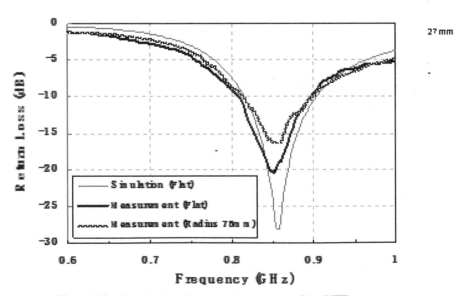

Figure 7.2   Simulated and measured return loss of the RFID tag antenna.

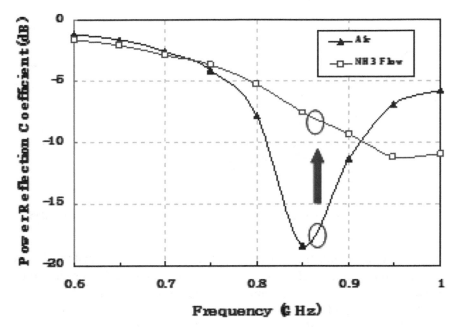

Figure 7.3 The calculated power reflection coefficient with and without ammonia.

Figure 7.3 shows the power reflection coefficient of the RFID tag antenna with SWCNT film before and after the gas flow.

Although the reported activity has to be considered pioneering because it testifies the feasibility of a fully organic, recyclable, passive and "multibit" RFID tag sensor, it is worth acknowledging that the exhibited dynamic range of about 10 dB is still not sufficient to read the presence of ammonia in a general application environment. Moreover, the field of application of such a system is limited by the fact that the conceived sensor encodes its information in the amplitude of the interrogating signal, which means that both the interrogating signal and the sensor reply occur at the same frequency and the system is not immune to the interference caused by reflections coming from the surrounding.

To overcome the limitations due to iso-frequency operation and the reduced dynamic range, a novel tag sensor architecture has been proposed, based on a frequency doubler and a Wheatstone bridge [24].

Figure 7.4 illustrates the architecture of the transponder.

On the one hand, the frequency doubler is aimed at separating the interrogating signal at f_0 and the sensor's reply at $2f_0$, thus solving the issue of the interferences between the transmitted and the received signals.

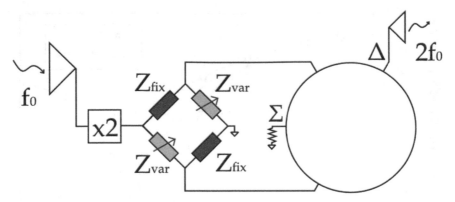

Figure 7.4 Block diagram of the harmonic sensor based on the Wheatstone bridge.

On the other hand, the sensing block consists of a Wheatstone bridge, where two opposite impedances are sensitive to changes of a certain parameter (for instance temperature, ammonia, and so forth), so that the bridge is forced to be in one of two states: balanced ($Z_{fix} = Z_{var}$), when no variation is sensed (quiescent mode), and unbalanced ($Z_{fix} \neq Z_{var}$), when a variation of the sensed parameter is experienced. From here on, the impedances of the bridge are considered to be real.

In quiescent condition, the two output signals from the bridge are equal in amplitude and phase. As a result, their difference, obtained thanks to a power combiner, is null and no signal is transmitted. When the bridge is unbalanced instead, the amplitudes of the output signals differ from one another, and the transmitted signal is proportional to the magnitude of the variation of the parameter under test.

To demonstrate the performance of the system, a complete prototype has been manufactured by using a paper substrate (see Figure 7.5) and a controlled experiment, conducted by using surface mounted resistances, has been performed.

The transponder is targeted for an $f_0 = 1.04$ GHz. The two fixed resistances are implemented by using $R_{fix} = 100\ \Omega$, whereas two different impedance values are considered for the variable ones: R_{var} is set equal to $100\ \Omega$ for the dubbed "balanced state" and equal to $25\ \Omega$ for the "fully unbalanced state". In this way, the dynamic range of the system in presence of an excursion of 75% in the value of the variable impedances is considered.

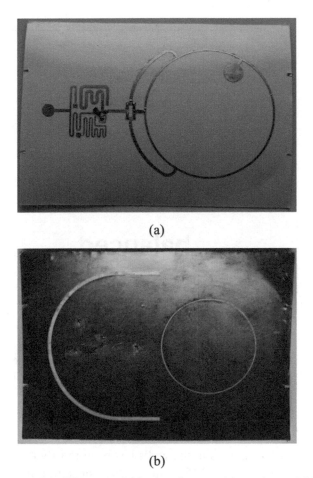

(a)

(b)

Figure 7.5 Layout of the Wheatstone bridge based sensor: (a) top view and (b) bottom view.

Figure 7.6 reports the measured received 2nd harmonic power versus tag-to-reader distance for a transponder in balanced and fully unbalanced condition.

The experiment is conducted with a transmitted power of 10 dBm EIRP and the receiver presents a noise floor around −100 dBm. A maximum excursion (actual dynamic range) of 55 dB is experienced at the distance of 10 cm, whereas it reduces for longer distances due to the presence of the noise floor. However, the two states are still identifiable at 50 cm, where the excursion is still on the order of 20 dB.

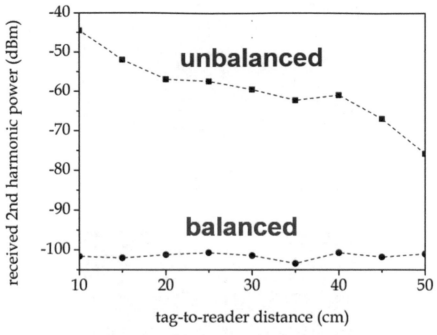

Figure 7.6　Measured received 2nd harmonic power versus tag-to-reader distance for a transponder in balanced and fully unbalanced condition; note that the reference balanced signal corresponds to the noise floor of the receiver adopted as a reader.

7.2.5 Zero-Power Wireless Crack Sensor

So far, the increase of cracks in the walls has been mostly detected by using small pieces of window glass [25], which represents a rudimentary approach not suitable for the implementation of real time monitoring of a large and dense population of buildings, since it is based on the visual inspection performed by a human operator. Other methods require the use of wired sensors, such as strain gages and accelerometers, which are usually impractical to install [26]. Current commercial wireless crack sensors are bulky, expensive and rely on batteries for their operation, which means that they need a periodical maintenance and are not suitable for massive deployment. As a consequence, recently, some studies have been conducted to fabricate wireless, compact and battery-less crack sensors which, however, suffer from low read range [27–29].

In order to overcome this possible drawback, an approach based on the harmonic radar principle has been proposed [30].

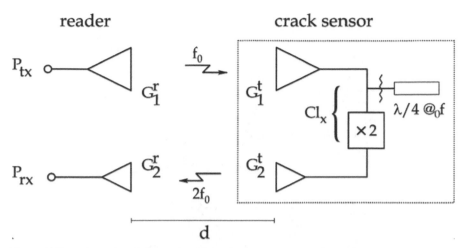

Figure 7.7 Schematic of the proposed crack sensor system. The main system parameters are: P_{tx} transmitted power at f_0; P_{rx} received power at $2f_0$; d distance; G_{ji} specific antenna gain (i = 1 fundamental, i = 2 2nd harmonic, j = t tag, j = r reader) and C_{lx} conversion loss (x = i intact condition, x = c cracked condition).

Figure 7.7 illustrates the block diagram of the proposed system. The reader interrogates the tag by sending a sinusoidal signal with a frequency f_0, called fundamental frequency. When the tag is intact (quiescent mode), the signal gathered by its input antenna is short-circuited by a stub placed at the input of the frequency doubler (which is a quarter-wave open-circuited stub designed at the fundamental frequency). As a consequence, the signal is prevented from entering the doubler and ideally no second harmonic is generated. On the other hand, if a crack occurs (alarm mode), the stub is turned off. In this case the signal can reach the doubler which converts it to $2f_0$ (second harmonic). The second harmonic is thus transmitted back to the reader and it is detected by the receiver, thus producing an alarm. The transponder is characterized by a known conversion loss C_l, defined as the ratio of the available power at f_0 at the input of the stub to the power delivered by the doubler to a 50 ohm load at $2f_0$. This amount depends on the tag operating mode and is also power-dependent (i.e. dependent on the interrogating distance).

Figure 7.8 shows a prototypal implementation of the crack sensor to prove the concept. The prototype is fabricated on a paper substrate by using the copper adhesive laminate technology [31]. The transponder is targeted for a fundamental frequency of 2.45 GHz and is based on a topology

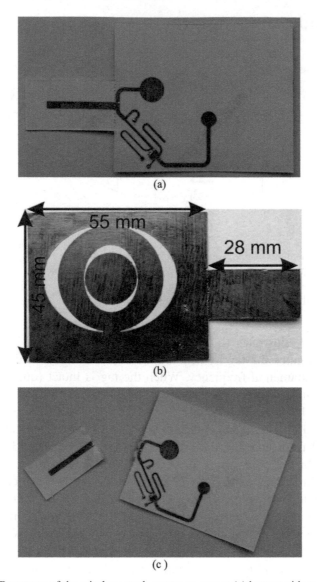

Figure 7.8   Prototype of the wireless crack sensor on paper: (a) bottom side and (b) top side in intact condition, and (c) cracked tag.

of nested tapered annular slot antennas and a zero-bias Schottky diode. The transponder is reported both in quiescent (Figure 7.8(a) and (b)) and alarm modes (Figure 7.8(c)).

The receiver can be implemented, for lab validation, by using a spectrum analyser. When the tag is intact the received signal is below the instrument noise floor, as reported in Figure 7.9, while, if the tag is cracked, a signal is detected at the second harmonic.

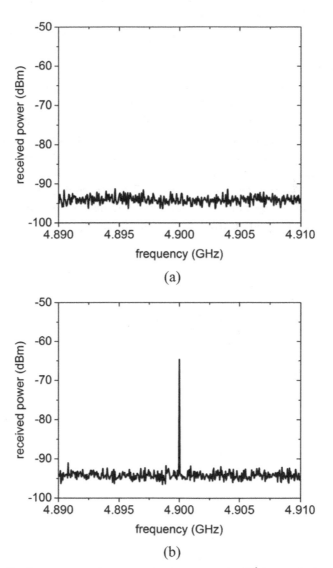

Figure 7.9 Crack sensor experiment: received power at the 2nd harmonic for an (a) intact and a (b) cracked transponder.

7.2.6 "Multi-Bit", Chip-Less Sensor Tag

A third possibility to transmit sensor data by means of a harmonic tag has been recently published [32] and uses the tag circuit of Figure 7.10. The basic idea is to encode the sensor information in the phase difference between two signals transmitted by two orthogonal antennas, one acting as the reference for the other one. In this way an accurate, relative measurement is possible.

To describe the tag operation let's consider the signal flow shown in the same figure. The incoming electromagnetic wave, at frequency f_0, is received by a spiral or a helical antenna. In this way the power at the antenna output is maximized, regardless of the polarization or the relative reader-tag orientation. The received power is then fed into a varactor or into a Schottky diode frequency doubler and the second harmonic is generated. At this point a power divider splits the $2f_0$ signal. The first half is directly re-irradiated in vertical polarization (e.g. E_η in Figure 7.10) in such a way as to form a reference signal component. The second half, instead, is phase-shifted by the angle Δ_Φ and then re-irradiated in horizontal polarization (e.g. E_ξ in Figure 7.10).

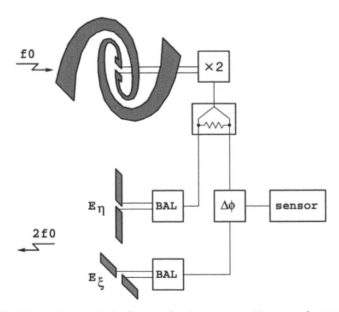

Figure 7.10   Harmonic tag block diagram for the system with two orthogonal antennas. The sensed quantity alters the phase of one channel, the other being used as a reference one [32, reproduced courtesy of The Electromagnetics Academy].

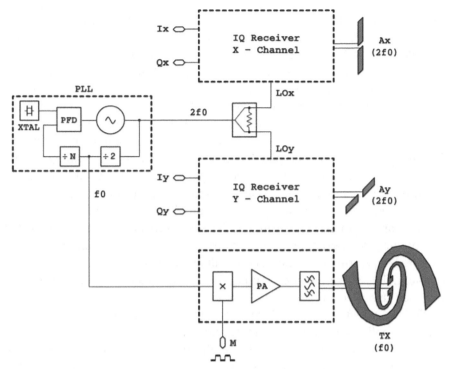

Figure 7.11 Reader block diagram for the system with two orthogonal antennas. Two I/Q receivers are used to retrieve the phase information [32, reproduced courtesy of The Electromagnetics Academy].

The phase angle Δ_Φ encodes the sensor information and must be recovered by the reader. To this purpose the reader is composed of four subsystems, as depicted in Figure 7.11. A PLL oscillator is used to generate both the f_0 and the $2f_0$ signals in a synchronous way. Then, two I/Q receivers can be used to retrieve the phase information.

Ultimately, with this approach, information sensed by the tag can be transferred to the reader without any carrier modulation, thus avoiding relevant electronics. Indeed, apart from the transducer, only a diode is required for frequency duplication.

7.2.7 Evolution Towards Distributed Sensing Enabled by WPT

Distributed sensing to its massive extent, can be seen as a specific evolution of the "smart dust" concept conceived at Berkeley [33] that, in turn, can be seen as an interpretation taken to the extreme of the WSN concept. "Smart dust" can be made theoretically possible if all the circuitry needed to perform the functionalities of a sensor node were shrunk until the node itself becomes invisible: energy harvesting, sensing, computation and interconnection are miniaturized to the infinitesimal. The concept is fascinating, visionary and actually has inspired many potential evolutions. Although not extreme as the original concept, distributed sensing can be considered as one of these evolutions inspired by the "smart dust", but characterized by a more realistic feasibility. The basic idea is to provide the environment with a plethora of sensing nodes, very simple and not necessarily as performing as the state of the art, able to detect a specific parameter (temperature, humidity, pressure, strain, smoke, gas, whatever) and transduce this parameter in an electrical one (voltage, current), so as to provide a signal that, after conditioning, can be stored or transmitted. The performance decrease due to the technological downgrading, implied by the extreme miniaturization, the ultra-low-power operating condition and the dramatic cost per unit reduction, can be potentially recovered by exploiting the enormous data redundancy and the increased spatial resolution of the system.

The so called "smart surface" concept can effectively summarize the potential applications of the distributed approach [34].

One of the current bottlenecks to the deployment of wireless sensor networks for distributed sensing is power supply of disposable sensing nodes. In fact, the periodical maintenance, entailed by the replacement of traditional batteries, cannot be effectively performed in systems consisting of a considerable number of hardly traceable nodes. Consequently, researchers are focusing on the challenging task of wirelessly recharging ubiquitous devices and machines to allow truly perpetual wireless-powered communication [35].

On the basis of the specific application, a distinction must be made in terms of energy requirements of the wireless nodes, which in turn determines the kind of energy sources exploited in their charging operation and the power management system.

On the one hand, there are wireless devices that require frequent or even continuous support at medium to high energy levels (see for instance mobile phones and other complex circuits). These devices demand the usage of intentional and reliable sources, which can be also used for battery-recharging operations.

On the other hand, there are ultra-low-power nodes, which need to be interrogated only few times a day, being usually off for the rest of the time. Due to their very low duty cycle and low power requirements, researchers are studying the opportunity to harness unintentional ambient energy sources, including RF signals for communications. However, typical ambient available electromagnetic (EM) power levels, generated from sources such as DTV, GSM and UMTS, hardly exceed 10.6–10.4 mW/cm^2, especially in indoor scenarios [36]. Such a low amount of available EM power is often not able to satisfy typical power management unit requirements, therefore inhibiting several practical applications. As a consequence, in some cases a trade-off approach, involving the deployment of intentional sources (so called "energy showers" [37], which periodically provide the requested low amount of energy to randomly distributed battery-less transponders, is advisable.

Regardless of the specific scenario and energy need, in order to allow for WPT, each wireless device in the network needs to be equipped with a rectifying antenna (or rectenna), i.e., a structure able to receive the incoming wireless signal and to convert it from the higher frequency used in the transmission to direct current (dc). Such a circuit consists of an antenna and a rectifier and is added to all the other RF circuitry needed for the communication [38]. Different architectures are possible for the transponder, resulting from a trade-off between performance and area consumption, and some examples from literature are here reported.

7.2.8 RF-Powered Environmentally-Friendly Transponders for Identification and Localization

This section describes the design of a novel WPT-enabled paper-based tag architecture for next-generation UWB-RFID systems developed within the GRETA project (GREen TAgs and sensors with ultra-wide-band identification and localization capabilities) [19]. This project is aimed at developing enabling technologies relying on environmentally friendly materials for distributed systems destined for identification, localization, tracking and monitoring in indoor scenarios.

In this context, tags are required to be:

- Localizable with sub-meter precision even in indoor scenarios or in presence of obstacles;
- Small-sized (maximum area in the order of a few square centimeters) and lightweight (i.e., without cumbersome batteries);
- Eco-compatible (i.e., made with recyclable materials);
- Conformable
- Low-cost to permit the deployment of several tags in the environment.

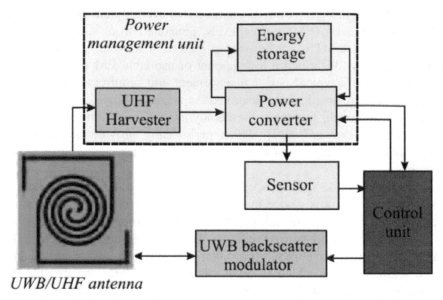

Figure 7.12 Schematic of an energy-autonomous UWB-RFID transponder.

The block diagram of the proposed transponder is reported in Figure 7.12. The system is designed for backscattering operations, in order to perform ultra-low power communication and localization by means of IR-UWB signals (European low UWB from 3.1 to 4.8 GHz) [36].

On the one hand, the UWB communication provides the tag identifier as well as the sensor data. On the other hand, energy harvesting and synchronization are ensured with the UHF link (868 MHz). The reader-tags synchronization can be performed by adopting a proper ASK or OOK modulation of the UHF carrier used for powering-up the tag. The UHF path is loaded by a rectifier and optimized in order to reach the highest RF-to-dc conversion efficiency. The obtained dc voltage is delivered to a power management unit in order to enable additional functionalities of the tag, e.g., sensing, or range extension for the UWB communication.

The dual-mode (UWB/UHF) operation is realized by using a low-profile paper-based UWB-UHF antenna and a miniaturized diplexer-like feeding network, located on the back side of the antenna (schematic reported in Figure 7.13).

The antenna design is based on the self-complementary architecture of an Archimedean spiral, which ensures almost constant radiating properties over the whole UWB. By extension of the spiral outer arms, a compact meandered

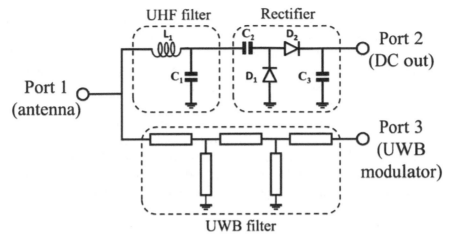

Figure 7.13 Circuit schematic of the three-port diplexer (UHF path connected to a full-wave rectifier and UWB signal connected to the backscatter modulator).

1.5-λ dipole is derived with resonance in the 868 MHz band. Besides size reduction, co-localization of the two radiating elements leads to a single-port antenna architecture, which is desirable for direct connection to future UWB-UHF integrated chips.

7.2.9 RF-Powered Implantable Sensors for Wireless Prosthesis Control

In this section an example of RF WPT enabling biomedical implantable wireless sensors is reported [39, 40].

The usage of batteries in implantable devices is highly not desirable for several reasons. First of all, the bulky size of batteries causes a substantial increase in the dimensions of the sensor node, which is unaffordable in most of the practical cases. Secondly, batteries have a limited life-cycle and battery replacement leads to surgical procedures that may cause infections. Moreover, the toxicity associated with the batteries is a main concern, since the sensors are in direct contact with the human tissue. Among the various alternative techniques to provide energy to a sensor, such as electromagnetic, thermoelectric, solar and motion energy scavenging (see [41–44]), radiative WPT is considered the most suitable for implantable devices, thanks to its ability to operate over relatively long distances and in presence of multiple nodes, and to the possibility to use controllable RF sources [45].

As an example, a system of sensors for targeted muscle re-innervation is considered, which are able to acquire EMG signals from re-innervated muscles and can allow for intuitive prosthesis control by an upper limb amputee. The reported RF-powered circuit, an ultralow-power SoC referred to as a "Bionode", is also optimized for multiple-node operation thanks to the setup of a synchronization system.

The block diagram of the proposed SoC is depicted in Figure 7.14. The system consists of six blocks: an RF power supply, a bio-sensing analogue front-end, a digital core, a process and voltage compensated clock oscillator, a receiver (RX) and a TDMA controller, and an ultralow power transmitter (TX) [39, 40]. The antenna topology is implemented by using electrically

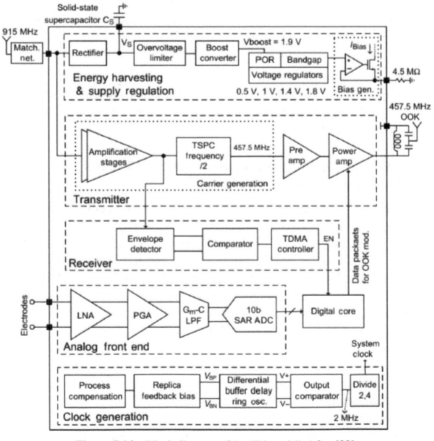

Figure 7.14   Block diagram of the "Bionode". After [39].

small printed meandered dipoles (off-chip). The impedance matching networks for both the rectifier input and the transmitter output are realized off-chip as well, but they could be merged into the respective antenna impedances in the full system.

All sensor nodes are powered by a base station mounted on the prosthetic arm, which broadcasts an RF signal at 915 MHz together with short-duration (9 μs) data used for synchronization.

A super capacitor is used to store the rectified RF energy. Then, the voltage VS at the capacitor, limited to a maximum of 0.8 V, is boosted by a boost converter up to 1.9 V. This voltage is utilized by the linear regulators to create four separate supply domains (0.5, 1, 1.4 and 1.8 V), whereas a bandgap reference and a bias current generation circuit are used to generate the reference voltages and bias currents. The system clock is provided by using a process voltage compensated 2 MHz on-chip oscillator [46].

Part of the powering RF signal is directly sent to the TX block as reference frequency. A frequency divider is used to generate the carrier frequency at 457.5 MHz and the obtained signal is modulated (OOK modulation) on the basis of the information provided by the analogue front-end block to the digital core. The data are transmitted to the base station, which is tuned to 457.5 MHz.

The digital core is controlled by a TDMA controller, which guarantees multi-node access. In particular, the RX block detects the short pulses associated with the RF powering signal by using an envelope detector and a comparator. Once a set number of pulses are counted by the TDMA controller, an "event" is triggered and the digital core allows for signal transmission.

Finally, the acquisition of the EMG signals is done by the analogue front-end, which includes a low-noise amplifier, a programmable gain amplifier, a lowpass filter, and an ultra-low-power ADC.

Figure 7.15 Figure shows the measured TDMA operation in the presence of three sensor nodes. A waveform generator is used in this case to create the trigger pulses for the RF signal generator.

The input RF signal is configured to have seven off-pulses in 8-μs period. The TDMA controller of each node counts the number of digital low signals in the aforementioned period and generates an "event" when the pulse count is between 4 and 8 to mitigate the impact of clock skew (in the zoomed graph we can see that the "node enable" signal for Node 2 rises after five off-pulses). When the "event" number matches with the hard-coded ID number of the

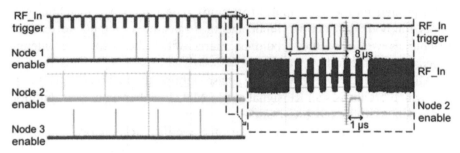

Figure 7.15 Measured time-domain TDMA synchronization control for three separate sensor nodes. After [39].

node an "enable signal" is generated for the digital core and the respective node transmits its own information.

Finally, a photograph of the chip is reported in Figure 7.16. It is implemented in a 1P6M 0.18 μm CMOS process and the overall die size is 1.35 mm × 1.5 mm.

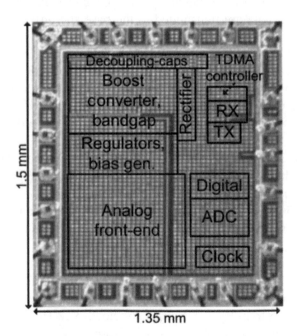

Figure 7.16   Die microphotograph. After [39].

7.2.10 RF-Power Temperature Sensor for Ambient Monitoring

The sensor described in the current section is aimed at performing the monitoring of environmental parameters, such as temperature, for IoT applications. In particular, the considered sensor is designed for the monitoring of food temperature, in order to guarantee the correct storage of food along its supply chain.

Figure 7.17 illustrates the block diagram of the reported temperature sensor node [47]. The receiver antenna, tuned to 2.4 GHz, is responsible for harvesting RF energy from the environment in the 2.4 GHz ISM band. The RF signal is delivered to a rectifier: a Dickson charge pump in CMOS technology optimized for low power levels. The generated dc signal is then stored in an external capacitor. The circuit stays in a standby status until the external capacitor has stored enough energy. The level of harvested energy is checked by the power management unit, which starts the measurement cycle once the required voltage across the capacitor is reached. A 100 kHz master clock generator, based on a current controlled ring oscillator, provides the time reference for all digital circuits of the sensor node.

The temperature is measured with an ultra-low power temperature to-digital sensor. This sensor codifies temperatures in the range between −10 to 30°C by using words of 8 bits. These words are sent to a modulator, which has the function of encoding the transmission frame containing the

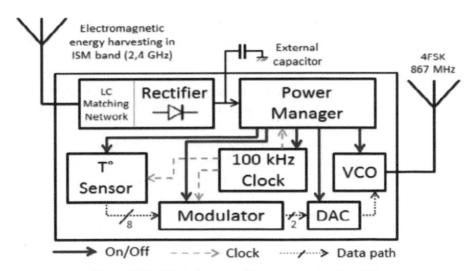

Figure 7.17   Block diagram of the sensor node. After [47].

raw measured data. The output of the modulator controls a VCO through a DAC. In particular, the DAC generates the four voltage levels needed by the VCO to yield the four frequencies of a 4-FSK signal. Finally, the TX antenna transmits the latter signal. The band of the re-transmitted signals lays around 867 MHz. Besides checking the charging level of the capacitor, the power management unit has the function of minimizing the power consumption by enabling/disabling each block.

7.3 IoS

7.3.1 The Eco-System

With respect to the IoT, where connectivity to people and things comes from Earth-based wireless networks, the IoS is based on connectivity from space-based satellites delivering information to every part of the world [48]. Also, IoS is recognized as a reference and opportunity for the development of satellite-based technology.

In recent years, there is a growing interest in space missions with small satellites in LEO. Small satellites (micro, nano, and pico class of spacecraft) have lower weights and smaller sizes with respect to the "traditional" artificial satellites, and are becoming more attractive due to lower development costs and shorter life times. Size, mass and power are an example of constraints for small satellite design and development, and therefore their functionality needs of miniaturization and integration technologies.

These classes of satellites will enable IoS with launching large number of satellites forming constellations or clusters, deployed as a network using inter-satellite connections enabling control, powering, communication and data processing with minimum human intervention. Such solutions are highly economical and future space missions will consist of multiple advanced, intelligent satellites that communicate with each other [49].

These systems have limited power, mass, antenna size, on-board resources, computing capabilities, etc. These features will foster the expansion of satellite networks with lower operational costs and with enhanced power management. Inter-satellite links, in fact, are feasible when satellite distances are small, considering also the system constraints as the limited amount of electrical power they can produce, the limited on-board power and computing resources.

7.3.2 The Future Trajectory of IoS

A crucial issue for the fostering of networks of small satellites in the IoS ecosystem is the power management. Moving towards technique such as WPT, relatively new for the space environment, can be a solution.

As well known, depending on the energy transfer mechanism, WPT can be classified into non-radiative and radiative categories. Non-radiative, or near-field power transfer, i.e. inductive (or capacitive) coupling between two separated coils, operates at short-range (distance less than the diameter of the transmitting coil) and at mid-range (distance varies from one to ten times the diameter of the transmitting coil) [50]. Radiative, or far-field power transfer technique uses the propagation of electromagnetic waves at long-range (distance of kilometres) by means of MPT and optical power transmission systems through high directivity antennas and laser beams, respectively, able to shape the reception area.

From this standpoint, WPT in the space environment is currently being studied from both the above two categories. The USA Department of Defense and NASA have started an investigation to provide an insight to develop a new control method for clusters of multiple satellites, and to demonstrate the ability of WPT across a mid-range distance, which could provide greater flexibility of future satellite cluster architectures [36]. This investigation (named DOD SPHERES-RINGS) uses two small satellites (SPHERES) fitted with donut-like rings (RINGS) to test wireless power transfer using electromagnetic fields. The RINGS hardware (aluminium coils and control systems) is placed around an individual satellite in order to demonstrate the use of electromagnetic coils to manoeuvre individual small satellites with respect to one another. The current running through the coils controls (attracts, repels and rotates) the satellites, and the next research goals will include improved control performance between separate satellites and the possibility of more efficient wireless power transfer at a greater distance than mid-range [51].

This is only an example of investigation steered towards an efficient way to transfer power among small satellites in order to reduce the need for alternative power sources. Further experiments in the WPT field will establish the hardware necessary for potential future powering of small satellites and enhanced communication systems in space (i.e. radiative power transfer, or far-field power transfer technique).

It is evident how the main challenge in WPT space applications is focused on far-field power transfer technique (long-range), with attention to the improvement of antenna beam directivity and efficiency [52, 53]. High gain antennas for MPT over kilometre distance with high efficiency can be used [52, 54], but some applications could require to transfer power using omnidirectional antennas to cover more areas. An advantage of MPT is that power can be transmitted to/from moving targets using beam-forming techniques, changing the beam direction using phased-array techniques. SPS is an example of space MPT, where satellites are equipped with large solar panels and convert the electricity generated by solar panels to high power microwaves beams (2.4 GHz to 5.8 GHz frequency range to achieved reasonable efficiencies) [52]. These beams are directed toward ground-based receiver stations having very large rectennas.

Therefore, since wireless MPT can link moving targets with high-beam efficiency [52], new perspectives can be opened and new applications can be investigated in the IoS eco-system. For instance, a large number of small satellites forming a constellation of clusters, arranged as a network, can be powered using only MPT delivered by a phased array installed on a master satellite equipped with solar panels (an idea can be deduced from Figure 7.18). The transmitted power can be received by rectenna arrays

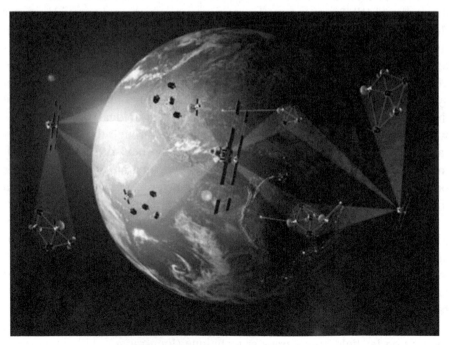

Figure 7.18   Inter-satellite links [55].

deployed on the micro/nano-satellite surface. Supplying microwave power to a network of orbiting platforms requires phased arrays with suitable efficiency and accurate target detection systems, and several investigations are being developed [52]. Trusting in new technologies and possible cost reduction of the employed microwave systems, such a IoS powering may be a future challenge to face.

For instance, a 3U CubeSat with dimensions of $10 \times 10 \times 30$ cm requires a power of the order of 10–50 W: considering the transmitting beam power of the master satellite of the order of kW, and depending on antenna distance (master-small satellite) and antenna aperture, the system could require phased array and rectenna with practicable efficiencies.

7.3.3 Satellite Cluster Vision

From the standpoint of the inter-satellite distances in a network, the MPT approach emerges as particularly suited for missions based on satellite clusters, the latter defined as a group of small satellites flying in a very close range (satellite distance 250 m–5 km) working to recreate the function of a single large satellite. A satellite cluster seems to be a new paradigm for space-borne surveillance and remote sensing [56]. For instance, these satellites can use synthetic aperture techniques to simulate a very large satellite with a very large actual antenna [57]. The cluster approach has many advantages with respect to a single large satellite:

- each spacecraft is small, light, simple to manufacture and relatively low cost;
- in the case of failure of a satellite, the failed spacecraft can be easily replaced;
- the cluster can reconfigure the satellite orbits to optimize for different missions.

A constellation of clusters would enable whole-earth coverage from low earth orbit and/or continuous coverage.

A first example of investigation in the formation of satellite cluster was the "TechSat 21" program (Technology Satellite of the 21st Century) [58], a coordinated study of several US Air Force Research Laboratories directorates to test technology for spacecraft, which can rapidly change formation based on mission requirements.

As possible remote sensing applications, the SAR imaging and passive radiometry observations were identified (Figure 7.19). Even though the

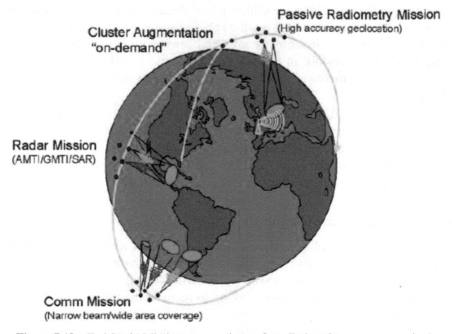

Figure 7.19   TechSat 21 Mission concept (image from Techsat 21 program overview).

project was cancelled in 2003 due to the budget limits, the new trajectory for cluster satellite missions has been traced.

The further vision of TechSat 21 was a constellation of "virtual" satellites in order to improve the Earth coverage, with each "virtual" satellite being a cluster of micro-satellites. For instance, it was hypothesized that each cluster would contain eight micro-satellites flying within 250 meters up to 5 km of each other, depending on the radar or radiometric application.

TechSat 21 foresaw, for each small satellite, to maintain sun pointing of solar arrays for power collection [58]: MPT technology could be the new vision for satellite cluster powering, restricting the transmission of a beam power to a master satellite equipped with large solar panels.

7.4 Conclusions

In the last decade WPT has become an increasingly popular technology in several areas of applications ranging from electric vehicle to consumer electronics, from RFID to long distance power transfer and even to solar space energy transfer to earth. Within this vast variety of applications, this chapter aimed at underlining how two more specific, different yet affine areas, namely IoT and IoS, can benefit from WPT development. In particular, the chapter shows, on the one hand the potentialities of WPT to provide solutions for problems posed by IoT in terms of increasing necessity to supply power to independent smart objects, thus allowing them to have the required sensing capabilities without either batteries or power grid connectivity. On the other hand, in a more visionary approach, the chapter also shows how WPT can foster satellite evolution towards clusters of cube-satellites in low earth orbit, thus enabling, among the others, further development of remote sensing potentialities.

References

[1] CISCO (2011). Available at: http://www.cisco.com/c/dam/en_us/about/ac79/docs/innov/IoT_IBSG_0411FINAL.pdf

[2] Phonesat the Smartphone Satellite (2013). Available at: https://www.nasa.gov/centers/ames/engineering/projects/phonesat.html

[3] Roselli, L., et al. (2016). "WPT related applications enabling internet of things evolution," in *Proceedings of the 10th European Conference on Antennas and Propagation,* Davos, 1–2.

[4] Ansari, F. (2005). *Sensing Issues in Civil Structural Health Monitoring.* Dordrecht: Springer.

[5] San-Miguel-Ayanz, S., et al. (2012). *Comprehensive Monitoring of Wildfires in Europe: The European Forest Fire Information System.* Brussels: European Commission.

[6] Heilig, A., Schober, M., Schneebeli, M., Fellin, W. (2008). "Next level for snow pack monitoring in real-time," in *Proceedings of the International Snow Science Workshop,* Fernie, BC.

[7] Goetz, S. J., Baccini, A., Laporte, N. T., Johns, T., Walker, W., Kellndorfer, J., et al. (2009). Mapping and monitoring carbon stocks with satellite observations: a comparison of methods. *Carbon Balance Manag.* 4:2.

[8] Wu, Y. M., and Zhao, L. (2006). Magnitude estimation using the first three second P-wave amplitude in earthquake early warning. *Geophys. Res. Lett.* 33, 1–16.

[9] Murvaya, P., and Silea, I. (2012). A survey on gas leak detection and localization techniques. *J. Loss Prev. Process Ind.* 25, 966–973.

[10] Acharjya, D. P., and Geetha, M. K. (2017). *Internet of Things: Novel Advances and Envisioned Applications.* Berlin: Springer.

[11] Anastasi, G., Farruggia, O., Lo Re, G., Ortolani, M. (2009). "Monitoring high-quality wine production using wireless sensor networks," in *Proceedings of the 42nd Hawaii International Conference System Sciences, 2009, HICSS '09,* Washington, DC, 1–7.

[12] Correia, R., Borges Carvalho N., and Kawasaki, S. (2016). Continuously power delivering for passive backscatter wireless sensor networks. *IEEE Trans. Microw. Theory Tech.* 64, 3723–3731.

[13] Karmakar, N. C. (2010). *Handbook of Smart Antennas for RFID System.* Hoboken, NJ: John Wiley & Sons.

[14] Nguyen, C. M., Mays, J., Plesa, D., Rao, S., Nguyen, M., and Chiao, J. C. (2015). "Wireless sensor nodes for environmental monitoring in Internet of Things," in *Proceedings of the 2015 IEEE MTT-S International Microwave Symposium,* Phoenix, AZ, 1–4.

[15] Lee, Y. H., et al. (2017). "Wireless power IoT system using polarization switch antenna as polling protocol for 5G mobile network," in *Proceedings of the 2017 IEEE Wireless Power Transfer Conference (WPTC),* Taipei, 1–3.

[16] Roselli, L., Mariotti, C., Virili, M., Alimenti, F., Orecchini, G., Palazzi, V., et al. (2016). "WPT related applications enabling internet of things evolution," in *Proceedings of the 10th European Conference on Antennas and Propagation,* Davos, 1–2.

[17] Rida, A., Yang, L., and Tentzeris, M. M. (2010). *RFID-Enabled Sensor Design and Applications (Artech House Integrated Microsystems).* Norwood, MA: Artech House.

[18] Finkenzeller, K., and Muller, D. (2010). *RFID Handbook: Fundamentals and Applications in Contactless Smart Cards, Radio Frequency Identification and Near-Field Communication,* 3rd Edn. Hoboken, NJ: Wiley.

[19] Decarli, N., Guerra, A., Guidi, F., Chiani, M., Dardari, D., Costanzo, A., et al. (2015). "The GRETA architecture for energy efficient radio identification and localization," in *Proceedings of the International EURASIP Workshop RFID Technology,* Oberaudorf, 1–8.

[20] Tedjini, S., Karmakar, N., Perret, E., Vena, A., Koswatta, R., and E-Azim, R. (2013). Hold the chips: chipless technology, an alternative technique for RFID. *IEEE Microw. Mag.* 14, 56–65.

[21] Kim, S., Mariotti, C., Alimenti, F., Mezzanotte, P., Georgiadis, A., Collado, A., et al. (2013). No battery required: perpetual rfid-enabled wireless sensors for cognitive intelligence applications. *IEEE Microw. Mag.* 14, 66–77.

[22] Roselli, L., Palazzi, V., Alimenti, F., Mezzanotte, P. (2017). "Towards multi-bit, long range and eco-friendly implementation of tag sensors," in *Proceedings of the 11th European Conference on Antennas and Propagation (EUCAP)*, Paris, 3922–3925.

[23] Yang, L., Orecchini, G., Shaker, G., Lee, H. S., and Tentzeris, M. M. (2010). "Battery-free RFID-enabled wireless sensors," in *Proceedings of the IEEE MTT-S International Microwave Symposium*, Anaheim, CA, 1528–1531.

[24] Palazzi, V., Alimenti, F., Mariotti, C., Virili, M., Orecchini, G., Roselli, L., et al. (2015). "Demonstration of a high dynamic range chipless RFID sensor in paper substrate based on the harmonic radar concept," in *Proceedings of the IEEE International Microwave Symposium (IMS)*, San Francisco, CA, 1–4.

[25] Monitoring and Evaluating Cracks in Masonry, U-S: General Services Administration (GSA) (2017). Available at: http://www.gsa.gov/portal/content/111626

[26] Humboldt Construction Materials Testing Equipmen (2017). Available at: http://www.humboldtmfg.com/concretecrack-monitor-crackgauge.html

[27] Yi, X., Wang, Y., Leon, R. T., Cooper, J., and Tentzeris, M. M. (2012). "Wireless crack sensing using an RFID-based folded patch antenna," in *Proceeding of the 6th International Conference on Bridge Maintenance, Safety and Management (IABMAS 2012)*, Lake Como, 8.

[28] Kalansuriya, P., Bhattacharyya, R., and Sarma, S. (2013). RFID tag antenna-based sensing for pervasive surface crack detection. *EEE Sens. J.* 13, 1564–1570.

[29] Kalansuriya, P., Bhattacharyya, R., Sarma, S., and Karmakar, N. (2012). "Towards chipless RFID-based sensing for pervasive surface crack detection," in *Proceedings of the IEEE International Conference on RFID–Technologies and Applications (RFID-TA)*, Warsaw, 5–7.

[30] Palazzi, V., Alimenti, F., Mezzanotte, P., Orecchini, G., and Roselli, L. (2017). "Zero-power, long-range, ultra low-cost harmonic wireless sensors for massively distributed monitoring of cracked walls," in *Proceedings of the IEEE MTT-S International Microwave Symposium (IMS)*, Honolulu, HI, 1–9.

[31] Alimenti, F., Mariotti, C., Palazzi, V., Virili, M., Orecchini, G., Mezzanotte, P., et al. (2015). Communication and sensing circuits on cellulose. *J. Low Power Electron. Appl.* 5, 51–164.

[32] Alimenti, F., and Roselli, L. (2013). Theory of zero-power RFID sensors based on harmonic generation and orthogonally polarized antennas. *Electromagn. Waves* 134, 337–357.

[33] Warneke, B., Last, M., Liebowitz, B., and Pister, K. S. J. (2001). Smart dust: communicating with a cubic-millimeter computer. *Computer* 34, 44–51.

[34] Roselli, L., et al. (2014). Smart surfaces: large area electronics systems for internet of things enabled by energy harvesting. *Proc. IEEE* 102, 1723–1746.

[35] Costanzo, A., and Masotti, D. (2017). Energizing 5G. *IEEE Microw. Mag.* 18, 125–136.

[36] Costanzo, A., Masotti, D., Fantuzzi, M., and Del Prete, M. (2017). Co-design strategies for energy-efficient UWB and UHF wireless sytems. *IEEE Trans. Microw. Theory Tech.* 65, 1852–1863.

[37] Masotti, D., and Costanzo, A. (2017). "Time-based RF showers for energy-aware power transmission," in *Proceedings of the 11th European Conference on Antennas and Propagation (EUCAP)*, Paris, 783–787.

[38] Yoo, T.-W., and Chang, K. (1992). Theoretical and experimental development of 10 and 35 GHz rectennas. *IEEE Trans. Microw. Theory Tech.* 40:6.

[39] Bhamra, H., Kim, Y.-J., Joseph, J., Lynch, J., Gall, O. Z., Mei, H., et al. A 24 μW, batteryless, crystal-free, multinodesynchronized SoC "bionode" for wireless prosthesis control. *IEEE J. Solid State Circ.* 50, 2714–2727.

[40] Kim, Y.-J., Bhamra, H. S., Joseph, J., and Irazoqui, P. P. (2015). An ultra-low-power RF energy-harvesting transceiver for multiple-node sensor application. *IEEE Trans. Circ. Syst.* 62:12.

[41] McDonnall, D., Hiatt, S., Smith, C., and Guillory, K. S. (2012). "Implantable multichannel wireless electromyography for prosthesis control," in *Proceedings of the Annual International Conferene IEEE EMBC*, Minneapolis, MN, 1350–1353.

[42] Jang, J., Berdy, D. F., Lee, J., Peroulis, D., and Jung, B. (2013). A wireless condition monitoring system powered by a sub-100 μW vibration energy harvester. *IEEE Trans. Circ. Syst. I Reg. Pap.* 60, 1082–1093.

[43] Guilar, N. J., Amirtharajah, R., and Hurst, J. P. (2009). A full-wave rectifier with integrated peak selection for multiple electrode piezoelectric energy harvesters. *IEEE J. Solid State Circ.* 44, 240–246.

[44] Chen, G., et al. (2010). Millimeter-scale nearly perpetual sensor system with stacked battery and solar cells. *Proc. IEEE ISSCC* 53, 288–289.

[45] Agarwal, K., Jegadeesan, R., Guo, Y.-X., and Thakor, N. V. (2017). Wireless power transfer strategies for implantable bioelectronics: methodological review. *IEEE Rev. Biomed. Eng.* 10:1.

[46] Bhamra, H., and Irazoqui, P. (2013). "A 2-MHz, process and voltage compensated clock oscillator for biomedical implantable SoC in 0.18um CMOS," in *Proceedings of the IEEE International Symposium Circuits Systems (ISCAS),* Montréal, QC, 618–621.

[47] Popov, G., Dualibe, F. C., Moeyaert, V., Ndungidi, P., Garcìa-Vàzquez, H., and Valderrama, C. (2016). "A 65-nm CMOS battery-less temperature sensor node for RF-powered wireless sensor networks," in *Proceedings of the IEEE Wireless Power Transfer Conference (WPTC),* Taipei.

[48] IEEE Internet of Things, The Internet of Space (IOS): A Future Backbone for the Internet of Things (2017). Available at: http://iot.ieee.org/newsletter/march-2016/the-internet-of-space-ios-a-future-backbone-for-the-internet-of-things.html

[49] Radhakrishnan, R., Edmonson, W. W., Afghah, F., Rodriguez-Osorio, R. M., Pinto, F., and Burleigh, S. C. (2016). Survey of inter-satellite communication for small satellite systems: physical layer to network layer view. *IEEE Commun. Surv. Tutor.* 18, 2442–2473.

[50] Shadid, R., Noghanian, S., and Nejadpak, A. (2016). "A literature survey of wireless power transfer," in *Proceedings of the 2016 IEEE International Conference on Electro Information Technology (EIT),* Grand Forks, ND, 0782–0787.

[51] Department of Defense Synchronized Position, Hold, Engage, Reorient, Experimental Satellites- Resonant Inductive Near-field Generation System (DOD SPHERES-RINGS) (2017). Available at: https://www.nasa.gov/mission_pages/station/research/experiments/916.html

[52] Shinohara, N. (2013). Beam control technologies with a high-efficiency phased array for microwave power transmission in Japan. *Proc. IEEE* 101, 1448–1463.

[53] Shinohara, N. (2010). Beam efficiency of wireless power transmission via radio waves from short range to long range. *J. Korean Inst. Electromagn. Eng. Sci.,* 10, 224–230.

[54] McSpadden, J., and Mankins, J. (2002). Space solar power programs and microwave wireless power transmission technology. *IEEE Microw. Mag.* 3, 46–57.

[55] I Links (2017). Available at: http://personal.ee.surrey.ac.uk/Personal/L. Wood/isl/

[56] Mohammed, J. L. (2001). Mission planning for a formation-flying satellite cluster. *Am. Assoc. Artif. Intell.*

[57] TECHSAT 21 and Revolutioning Space Missions using Microsatellites. Air Force Research Laboratories Space Vehicles Directorate (1998). Available at: http://www.kirtland.af.mil/Units/AFRL-Space-Vehicles-Directorate/

[58] Burns, R., et al. (2000). TechSat 21: formation design, control, and simulation. *IEEE Aerosp. Conf. Proc.* 7, 19–25.

8

IoT

Ricardo Correia, Daniel Belo and Nuno Borges Carvalho

Instituto de Telecomunicações, Departamento de Electrónica,
Telecomunicações e Informática, Universidade de Aveiro, Aveiro, Portugal

8.1 Introduction

The rapid increase in the progress and development of wireless communications and identification made it possible to track and sense some materials wirelessly. Nowadays, the continuously growth of connected objects will led to billions of wireless sensing devices in the Internet of Things (IoT) grid, up to 50 billion by 2020. IoT is transforming everyday physical objects that surround us into an ecosystem of information that will enrich our lives. While the IoT represents the convergence of advances in miniaturization, wireless connectivity, increased data storage capacity and batteries, the IoT would not be possible without sensors. The growth of the devices will be made possible only if the sensors battery needs are eliminated or reduced significantly. For low-power sensors and devices, careful power management and power conservation are critical to device lifetime and effectiveness. One of the possible solutions is to change completely the paradigm of the radio transceivers in the wireless nodes of the IoT system. The new paradigm should be able to communicate and possibly to power up sensors using only electromagnetic waveforms transmitted over the air. In this context, the frequent battery maintenance of all wireless nodes is undesirable or even impossible and passive backscatter radios will have an important role due to their low cost, low complexity and battery-free operation.

The goal of IoT is to provide connectivity to everything, anywhere, anytime. Since the conventional technologies like ZigBee, Bluetooth and Wi-Fi have high power consumption levels (from 10 mW to 1000 mW as can be seen in Figure 8.1), low profile technologies featuring low power consumption, ideally battery-free, low cost and low complexity, are necessary

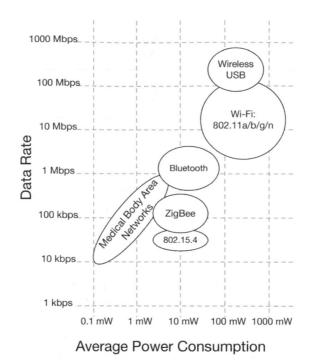

Average Power Consumption

Figure 8.1   Power consumption as function of data rate of different technologies.

to allow ubiquitous applications. Therefore, there is a huge need to design novel wireless communication techniques to achieve higher data rates while simultaneously minimising energy consumption.

8.1.1 Backscatter Communication

The difference between passive and active RFID wireless transceivers is the backscatter modulation [1] for the uplink. In the backscatter communication, shown in Figure 8.2, the tag reflects a radio signal transmitted by the reader, and modulates the reflection by controlling its own reflection coefficient [2]. The load modulator is usually a transistor switch that changes between two different impedances. By switching the antenna impedance between two values, the tag can binary modulate the RF signal that is scattered back to the reader. This communication is the reason for no need of active components found in traditional wireless transceivers designs, and it is the basis of the low-power implementation. Therefore, each sensor implements backscatter

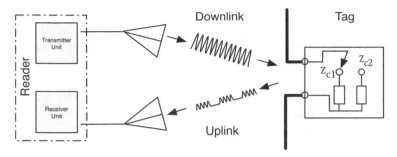

Figure 8.2 Schematic of backscatter communication.

communication by simply switching on and off the impedance connected to the antenna port.

The backscatter reader sends a continuous carrier wave which travels through a certain distance and reaches the backscatter device. A portion of the carrier power is converted by the backscatter device into dc current and is stored locally as harvested energy. The other portion of the carrier is reflected back to the reader by simply switching ON and OFF an RF transistor. These variations can be detected by the reader and decoded as the information transmitted by the backscatter device. The backscatter tags, by receiving the local oscillator over the air, do not need their own RF oscillator or phase-locked loop (PLL). By removing these circuits one reduces the power consumption and the cost of a tag chip [3]. Nevertheless, the backscattered signal received by the readers is weak, thus limiting communication range. In conventional backscatter communication, the tags must harvest enough power from the reader to turn on and modulate data, and the readers must receive strong backscattered signal to operate.

The modulation schemes frequently used are ASK and PSK. Some researchers have demonstrated other higher-order modulations schemes [4, 5], but due to the low SNR at the receiver, the most common are the low-order modulations. Some work on the characterization of the backscatter modulation has been developed in [6], where the impedance of a backscatter load modulator was measured to properly characterize the impedance matching of the RFID tags.

An energy harvester and power management circuitry are responsible for collecting sufficient energy to power the tag and any additional sensor. As previously mentioned, this energy can come from a variety of sources but is typically reader-delivered RF power. Over the last years, RF energy harvesting, which is the capability of converting RF signals into electricity, has gained a lot of interest [7]. This renewed interest for the area already resulted in the first commercial solutions. With the growth of portable applications, many companies, such as eCoupled, WiPower, and Powermat have developed solutions for the commercial market of wireless energy transfer. Thus, there is a strong motivation to enable a WSN with WPT which could be capable to afford all the operational costs.

In a backscatter RFID system, it is essential to ensure a power efficient communication. If the power at the tag's chip is smaller than the tag's sensitivity (minimum power required to power up the tag), the backscatter communication system is limited in its forward link. Nevertheless, if more than the ID is needed, or if it is intended to extend the range of backscatter radio, a battery is needed.

Some recent work [8, 9] show the possibility of using one frequency to continuously power the wireless sensor and other to transfer data by backscatter means with ASK modulation. Figure 8.3 presents a potential solution presented in [9] for this implementation.

As illustrated in Figure 8.4, without the need to replace energy depleted sensors in conventional WSNs, a passive WPT WSN can achieve a continuous operation with a large number of sensors powered by fixed energy transmitters used for both wireless charging and data collection. Nevertheless, due to the more ample power supply from WPT, RFID devices can

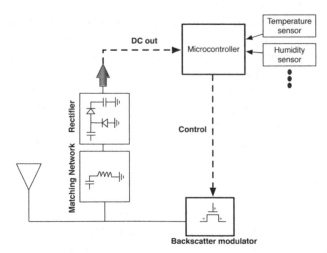

Figure 8.3   Block diagram of the solution presented in [8].

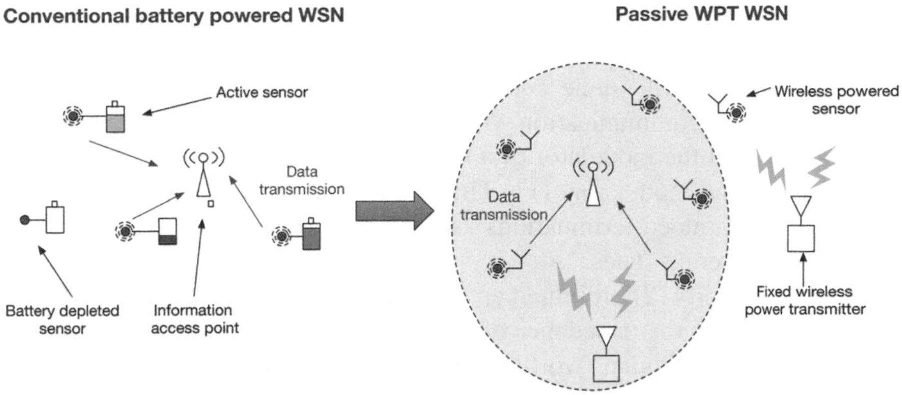

Conventional battery powered WSN

Passive WPT WSN

Figure 8.4 Example applications of a conventional battery powered WSN and a passive WPT WSN system. The red lines represent information, and the green represent energy.

now expect much longer operating lifetimes, and afford to transmit actively at a much larger data rate and from a longer distance than conventional backscatter based RFID communications. Therefore, we envision that a WSN with WPT will be an important building block of many popular commercial and industrial systems in the future, including the upcoming Internet of Things/Everything (IoT/IoE) systems consisting of millions of sensing/RFID devices as well as large-scale WSNs.

8.1.1.1 High data rate backscatter QAM modulation

In most RFID systems and passive sensors, the reader to tag communication is an ASK or PSK that modulates either the amplitude, or both the amplitude and phase, of the reader's transmitted RF carrier. The use of this technology entails a number of advantages over barcode technologies such as tracking people, items, and equipment in real time, non-line of sight requirement, long reading range, and standing harsh environment. However, a recent work [10] has shown that modulated backscatter can be extended to include higher order modulation schemes, such as 4-QAM. While ASK and PSK transmit 1 bit of data per symbol period, 4-QAM based can transmit 2 bits per symbol period, thus increasing the data rate and leading to reduced on-chip power consumption and extended read range.

The work presented in [4, 10] refers to a 4-QAM backscatter in semipassive systems, by using a coin cell battery as a power source for the modulator and a microcontroller that needs 3 V of supply. This way, the authors proved the QPSK modulator and battery powered system, by using an approach with a four lumped impedances connected to a RF switch that is controlled by a microcontroller. The same authors developed a 16-QAM modulator for UHF backscatter communication with a consumption of 1.49 mW at a rate of 96 Mbps only in the modulator (not the overall system with data generation logic feeding the modulator) [11].This modulator was implemented with 5 switches with lumped terminations as a 16-to-1 Mux to modulate the load between 16 different states.

The authors in [12] presented an I/Q backscatter modulator that use bias currents to change the impedance of two PIN diodes. The circuit comprises a Wilkinson power divider, two filters (low pass and high pass) to guarantee symmetrical paths on the board, one in each branch, and one PIN diode for each branch. The bias consumption of the circuit is 80 mW (excluding DACS and FPGA logic), which imposes high power consumption and data rate limitations. The use of this circuit for a low power sensor is not feasible. Another approach by using the PIN diodes was presented in [13]. Pozar presented a reflection type phase shifter using a quadrature hybrid and two PIN diodes. By biasing the diodes to the ON and OFF state it was possible to change the total path length for both reflected waves, producing a phase shift at the output. However, the consumption of the PIN diodes is not suitable for low power sensors.

One different solution to obtain a BPSK modulation was presented in [14]. The authors presented a phase-shift modulator with two switches that are connected to each other by a 90°delay line or 0°delay line. The phase-shift modulator was implemented as a two-port device that selectively delays the signal by 90°between port 1 and port 2, or passes the signal from port 1 to port 2 with no delay, achieving a BPSK modulation.

In [15], two multi-antenna technologies were used. The authors presented an energy harvester (staggered-pattern charge collector (SPCC)) that has two independent antenna arrays and harvests power to supply a microcontroller. They also present a retrodirective array phase modulator (RAPM) that backscatters the signal from the reader to the reader. The RAPM comprises two switches that are controlled by a microcontroller. The four switching states are connected through coplanar waveguides of different wavelengths, each separated by 90°. The RAPM can be used for QPSK modulation by calibrating the phase offsets in the switches. The solution presented in our manuscript is more compact and more capable to be expanded to higher modulation orders when compared with the one presented in [15].

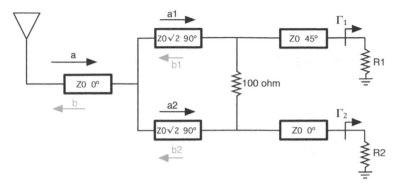

Figure 8.5 Model of the QAM solution.

In order to minimize the dimensions of the systems analyzed and to increase the order to a 16-QAM solution, the implementation on Figure 8.5 was evaluated. This model is constituted by a Wilkinson power divider and two different branches, each one terminated with a line and an ideal impedance. The lines present a 45°phase shift with each other, meaning in a 90°phase difference in the reflected wave of each branch. By varying the ideal impedance between two values it is possible to achieve a 4-QAM solution and by varying the impedance between 4 values it is possible to scale up the order to 16-QAM.

Figure 8.6 illustrates the diferences of line length in each branch, which are related to 45°phase shift. By using this approach, the circuit in Figure 8.6 was designed and each transistor is switched for different voltage levels to achieve different reflection coefficients.

Figure 8.7 presents the photograph of measurement setup used to acquire the reflection coefficient from the proposed circuit, illustrated in Figure 8.6.

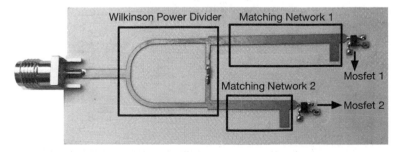

Figure 8.6 Photograph of the QAM backscatter circuit.

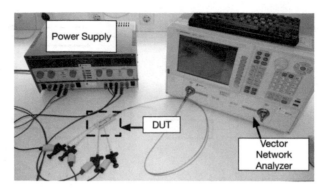

Figure 8.7   Photograph of measurement setup.

A power supply and a Performance Network Analyzer (PNA) (E8361C, from Agilent Technologies) was used, and was calibrated at 0 dBm for the frequencies from 2 GHz to 3 GHz. The results of the simulations and measurements are presented in Figure 8.8. The switches were changed according to the simulated results, with different voltage levels at each gate of transistor.

In order to give more precise and accurate results in real scenario applications, some simulations were conducted for different input power levels. In Figure 8.9c it is possible to analyze that for higher input levels the constellation will be more difficult to implement. For lower input powers it

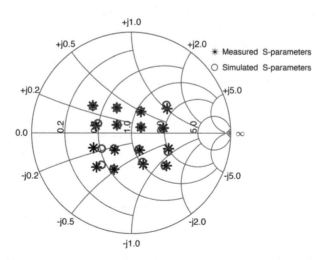

Figure 8.8   Simulated and measured S_{11} for different voltage levels at the gate of each transistor for 2.45 GHz at 0 dBm.

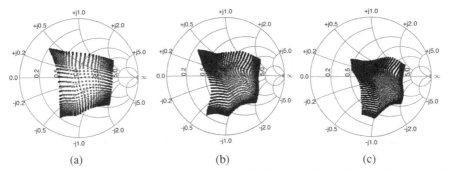

Figure 8.9 Simulated S11 for different voltage levels (from 0 V to 0.6 V with a step of 0.01 V) at the gate of each transistor for 2.45 GHz (a) Pin = –10 dBm. (b) Pin = 0 dBm. (c) Pin = 5 dBm.

is possible to obtain 16-QAM modulation as can be seen in Figure 8.9a and Figure 8.9b. In order to improve the results for higher input power levels, the circuit should be optimized for that levels.

The maximum data rate of the communication on the modulator link was studied and evaluated by using the measurement setup presented in Figure 8.10. This measurement setup was used to view the received constellation at multiple transmission rates. A vector signal generator (ROHDE & SCHWARZ SMJ 100A) to generate the transmitter signal at 2.45 GHz was used. An arbitrary waveform generator (TEKTRONIX AWG5012C) was used to generate the different voltage levels at the gate of each transistor.

Different voltage levels were applied into the gate of each transistor and 16 different impedances were generated and analyzed in the signal &

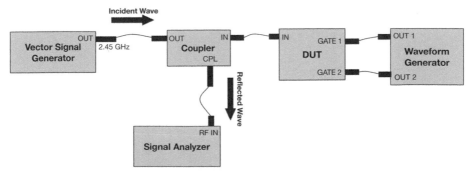

Figure 8.10 Block diagram of measurement setup for demodulation and achievable data rates.

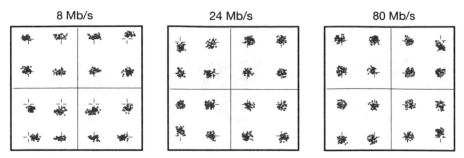

Figure 8.11   Received 16-QAM constellation for a center frequency of 2.45 GHz for different data rates.

spectrum analyzer. The received constellations are shown in Figure 8.11 for bit rates of 8 Mb/s, 24 Mb/s and 80 Mb/s. The constellation points lie in the appropriate quadrants, and each point is clearly visible. In order to perform better results, the system can be calibrated by changing the gate voltage level.

Figure 8.12 illustrates the measured EVM for different data rates. As the data rate increases, the measured EVM also increases. Nevertheless and despite the EVM seems high, the process has not passed through any signal processing equalisation at the receiver. These results clearly state the viability of the proposed solution for a high bit rate backscatter approach.

In Figure 8.13, the measured EVM of the 16-QAM modulator as function of input power for different data rates can be seen. Since this device was optimized for a 0 dBm (1 mW) RF input signal, the lowest value of EVM can be seen to be around this RF power. As can be seen from the figure, for a data rate of 120 Mb/s the system EVM is 16.7 %, for a data rate of 60 Mb/s

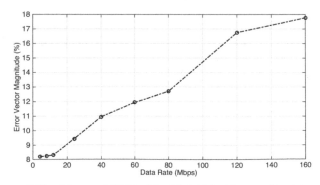

Figure 8.12   Measured EVM of the 16-QAM with a varying data rate.

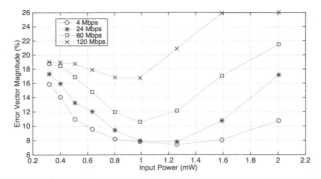

Figure 8.13 Measured EVM of the 16-QAM modulator as function of input power for different data rates.

the system EVM is 10.3%. The viability for this type of solution could be improved for lower values of power also, if the circuit was tuned for lower values of input RF power.

8.1.1.2 Backscatter QAM with WPT capabilities

Figure 8.14 shows block diagram of a system which combines the QAM backscatter modulator with WPT. This solution presents a potential performance improvement when compared with conventional battery-powered

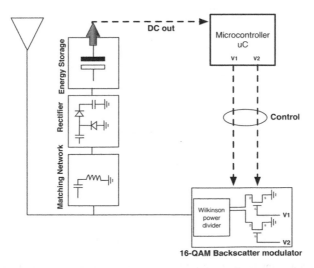

Figure 8.14 Block diagram of WPT with 16-QAM backscatter modulator.

wireless sensors network, because it eliminates the need of battery replacement or recharging. Using WPT reduces the operational cost and increase the communication performance. Since the wireless sensors need to harvest enough energy before sending data, with this solution it is possible to overcome these limitations by achieving continuous power delivery and increase the communication performance. The solution uses two different frequencies, one for WPT and the other for the QAM backscatter modulation and therefore it will have an important role in many popular commercial and industrial systems in the future of IoT systems.

The system proposed in Figure 8.14 can be divided into two main blocks, the 16-QAM backscatter modulator and the rectifier. Figure 8.15 presents the photograph of the proposed system. The first, which is the modulator, is composed by a Wilkinson power divider, two matching networks and two E-pHEMT transistors (Broadcom ATF-54143), following the same approach used in [16]. It was also proposed a five-stage Dickson multiplier to harvest the maximum dc power from the CW of transmitter. RF Schottky diodes (Skyworks SMS7630-006LF) were used and the rectifier was optimized for 0 dBm at 2.45 GHz.

The achievable data rates with the system presented in Figure 8.14 were evaluated by using the same setup shown in Figure 8.10. The modulation

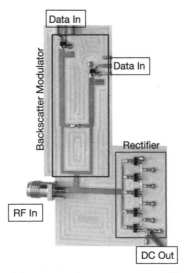

Figure 8.15   Photograph of the proposed system, composed by a 16-QAM modulator and a rectifier.

is performed by changing the transistor's gate voltage which changes the drain impedance and causes different reflection coefficients. The received constellations as well as the EVM and power consumption per bit as function of input power for 4 Mb/s, 400 Mb/s and 960 Mb/s were analyzed by signal & spectrum analyzer and are presented in Figure 8.16. Figure 8.16c presents an EVM of 12.8% corresponding to 0.33 pJ per bit of energy consumption for a data rate of 960 Mb/s. The EVM can decrease to 6.2% for a data rate of 4 Mb/s with 0.2 pJ per bit of energy consumption as can be seen in Figure 8.16a.

The received eye diagrams for varying data rates are shown in Figure 8.17. For the data rate of 4 Mb/s the eye is clearly open and the received signals are clean. However, if the data rate increases up to 960 Mb/s the eye becomes to close and the received signals present some noise. Nevertheless, it is possible to demodulate at such higher rates as it was seen from Figure 8.16.

Figure 8.15 shows two main blocks, the backscatter modulator and the rectifier. The efficiency of the five-stage rectifier is very low (around 6% for an input power of 5 dBm). To improve these results, some optimization can be performed or even an approach similar as [9] using two different frequencies. Nevertheless, it is possible to harvest some amount of dc power from the transmitter while communication is performed. The results of the collected dc output voltage of the system as function of input power are provided in Figure 8.18.

With all these features this system is suitable for providing high-bandwidth for future low power devices, such as IoT sensors.

8.1.1.3 Efficient wireless power transfer system for a moving passive backscatter sensor

In this subsection, a CW energy efficient transmitter working at 5.8 GHz will be presented, as a possible solution to realize an efficient wireless power transfer system, from the energy consumption at the transmitter, to the dc energy produced at an IoT sensor which is similar to the ones previously presented [12]. It is shown that the transmitting antenna is able to switch between three different states, with different antenna gain and radiated powers, based on a feedback signal that is provided by a pilot signal that can be backscattered from the IoT sensor at 3.45 GHz. The aim of this approach is to power up a moving IoT sensor. The single antenna element is depicted on Fig 8.19 a). The antenna uses a double layer aperture coupling technique and it is shown that the feed for each element is composed by a Single-Pole Double-Trough (SPDT) switch, where one of its outputs is connected to a power amplifier that drives the patch and the other to a 50 Ohm load.

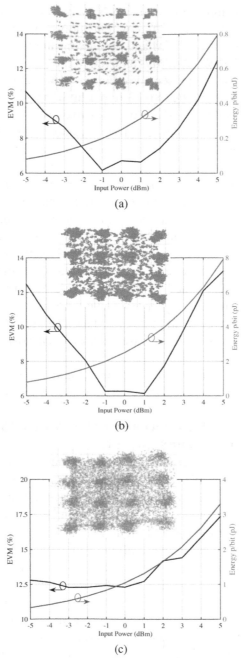

Figure 8.16 Received constellations with EVM and energy per bit consumption as function of input power for different data rates. (a) 4 Mb/s. (b) 400 Mb/s. (c) 960 Mb/s.

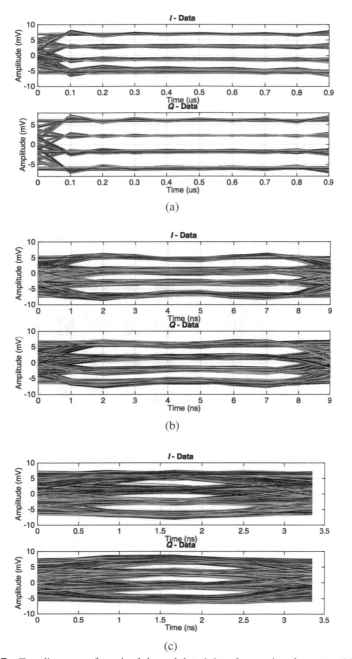

Figure 8.17 Eye diagrams of received demodulated data for varying data rates. (a) 4 Mb/s. (b) 400 Mb/s. (c) 960 Mb/s.

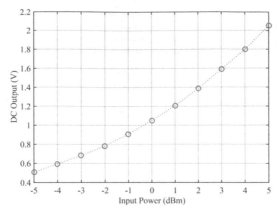

Figure 8.18   Experimental dc output voltage of system as function of input power.

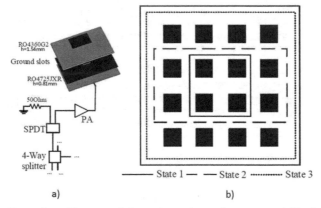

Figure 8.19   Reconfigurable transmitting microstrip patch antenna. a) Single antenna element configuration. b) Illustration of the active antenna elements that compose the three antenna states, on a 4x4 array.

The antenna has only one input and the splitting is made by means of 4-way signal splitters. In order to reconfigure the 4x4 antenna into the states shown in Figure 8.20 b), a 4-bit decoder is used to easily switch the SPDTs. Note that when the switches are configured to route the signal to the 50 Ohm loads, their respective PAs are at an off-state, allowing to save power. A single patch antenna designed for 3.45 GHz is placed side by side of this one in order to radiate the pilot signal that will be absorbed or reflected by the IoT sensor.

These three states will have different input matches, different gains, and different radiated powers based on how many antenna elements are active at

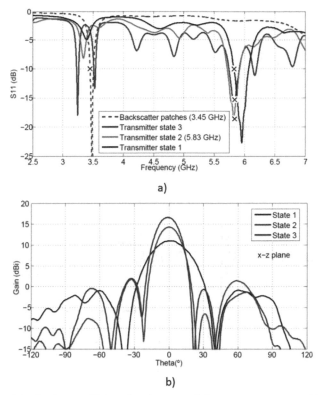

Figure 8.20 a) Reflection coefficient for the transmitting antenna states and backscattering patches b) Measured gain and radiation patterns of the three state transmitting antenna.

each time instant. Figure 8.20 a), presents the reflection coefficient of this antenna for the three states, as well as for the backscattering pilot signal patches and Figure 8.20 b) the gain of each state. Note that, if one say that the radiated power at state 1 is $Prad$, then the radiated power at state 2 is $Prad + 3\,dB$ and at state 3, $Prad + 6\,dB$. This approach allow us to operate the PAs either on an off-state or on their saturation, maximizing the efficiency of the transmitter.

To be able to continuously receive energy and simultaneously inform the transmitter about its operating RF-dc conversion efficiency value, a backscattering system was attached to the rectifier as shown in Figure 8.21 a). The circuit was optimized in such a way that the application of a dc voltage at the switching transistor, makes the backscattering patch to reflect back to the transmitter, the pilot signal at 3.45 GHz. The rectifier itself, is a single series

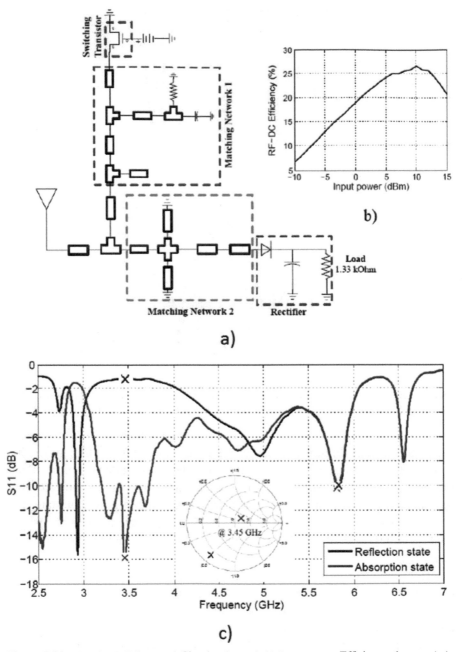

Figure 8.21 Passive IoT Sensor a) Circuit schematic b) Input power-Efficiency characteristic curve and c) Sensor reflection coefficient in both reflective and absorption states.

Schottky diode that is matched to 5.83 GHz, either if the transistor is on the on-state or off-state.

In order to provide the voltage needed to switch the transistor, a sample of the dc voltage achieved by the rectifier is used. Therefore, the voltage that is pre specified in the design to turn on the transistor, corresponds to a priori known efficiency operating point that can be derived from 8.1.

$$\eta = \frac{P_{dc}}{P_{in,RF}} = \frac{V_{dc}{}^2}{R_L \times P_{in,RF}}, \tag{8.1}$$

where $P_{in,RF}$ is the average input power at the sensor, P_{dc} the output dc power, V_{dc} the output dc voltage and R_L the load of the sensor. This way, the transmitter may change its state based on the reflection of the pilot signal due to the transistor on-state.

Referring to Figure 8.21 b), where the input power-efficiency curve of this specific receiver is represented, it is possible to identify a point where the maximum RF-dc conversion efficiency occurs. For lower input powers, the forward threshold level of the diode is the limiting factor and for higher powers, the breakdown region is the limiting factor. Note that this curve is a general RF-dc conversion efficiency curve. Taking this into account, by activating efficiently the antenna states it is possible to keep the receiver on a pre-determined efficiency operating point. By combining the concepts introduced in the previous sections, it is possible to maintain the IoT sensor operating at a certain specified efficiency point, while keeping the transmitter also at its maximum efficiency, for a CW signal.

In order to validate the approach, an experiment consisting on moving an IoT sensor on a straight line aligned with the transmitting antenna was performed and the setup is shown in Figure 8.22 a). On the Figure 8.22 b), both wireless power transfer and backscatter antennas are combined with a power combined and a directional coupler is used to monitor the input power. The three curves represented in Figure 8.22 c) are the efficiency distributions of the receiver over the covered distance, for each transmitter state. The experiment was conducted on a real laboratory environment where multipath effects can be easily found. By switching on/off the antenna elements and their respective PAs, it is possible to delivery energy efficiently at the IoT sensor, as shown. For a given input power at the antenna connector, P_{in}, the power radiated by the antenna is passively controlled by the selection of the states, depending on the distance at which the IoT sensor is by monitoring the backscattered pilot signal. Moreover, the antenna gain itself also changes, contributing also to an efficient overall coverage.

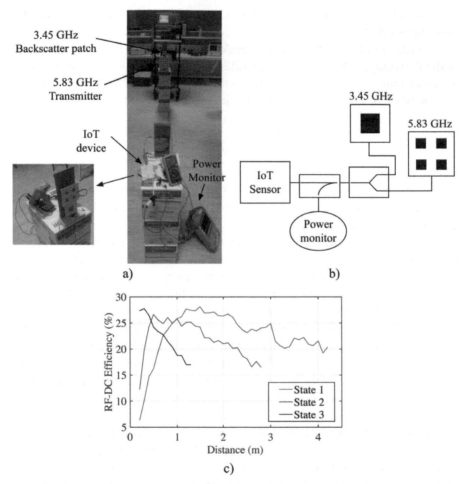

Figure 8.22   Passive IoT Sensor a) Circuit schematic b) Input power-Efficiency characteristic curve and c) Sensor reflection coefficient in both reflective and absorption states.

The power amplifier is the component whose efficiency dominates the overall efficiency of the transmitter and, by using a CW signal, those can be operated in saturation at all times. Assuming that the IoT sensor must operate with a minimum 20% efficiency at all, then the antenna state 1 would cover up to 1 m, then it would switch by monitoring the backscattered signal to state 2, with double power and cover up to 2.2 m and finally, state 3, that further increases the transmission power to double and cover the remaining distance, up to 4 m. It must be noted that, at the IoT sensor, a power combiner was used to combine both wireless power and backscattering signals collected by the antennas. Additionally, a directional coupler was inserted for input power monitoring (Refer to Figure 8.22 b)). By adding this components, the losses introduced by them contributed to a decrease of the covered distance.

References

[1] Fuschini, F., Piersanti, C., Paolazzi, F., and Falciasecca, G. (2008). Analytical approach to the backscattering from UHF RFID transponder. *IEEE Antennas Wirel. Propag. Lett.* 7, 33–35.
[2] Griffin, J. D., and Durgin, G. D. (2009). Complete link budgets for backscatter-radio and RFID systems. *IEEE Antennas Propag. Mag.* 51, 11–25.
[3] Daniel, K., and Popovic, Z. (2013). How good is your tag? RFID backscatter metrics and measurements. *IEEE Microw. Mag.* 14, 47–55.
[4] Stewart, T., and Reynolds, M. S. (2010). "QAM backscatter for passive UHF RFID tags," in *Proceedings of the 2010 IEEE International Conference on RFID (IEEE RFID 2010)*, (Warsaw: IEEE), 210–214.
[5] Correia, R., and Borges Carvalho, N. (2016). "Design of high order modulation backscatter wireless sensor for passive IoT solutions," in *Proceedings of the 2016 IEEE Wireless Power Transfer Conference (WPTC)*, Jersy: IEEE, 1–3.
[6] Marina, J., Correia, R., Ribeiro, D., Cruz, P., and Carvalho, N. B. (2016). "RF-to-DC and backscatter load modulator characterization," in *Proceedings of the 2016 87th ARFTG Microwave Measurement Conference (ARFTG)*, (San Francisco, CA: IEEE), 1–4.
[7] Visser, H. J., and Vullers, R. J. M. (2013). RF energy harvesting and transport for wireless sensor network applications: principles and requirements. *Proc. IEEE* 101, 1410–1423.
[8] Correia, R., Borges De Carvalho, N., Fukuda, G., Miyaji, A., and Kawasaki, S.-G. (2015). "Backscatter wireless sensor network with WPT capabilities," in *Proceedings of the International Microwave Symposium*, (San Francisco, CA: IEEE), 1–4.
[9] Correia, R., Borges Carvalho, N., and Kawasaki, S. (2016). "Continuously power delivering for passive backscatter wireless sensor networks," in *Proceedings of the IEEE Transactions on Microwave Theory and Techniques*, (Nanjing: IEEE), 1–9.

[10] Stewart, J. T., Wheeler, E., Teizer, J., and Reynolds, M. S. (2012). Quadrature amplitude modulated backscatter in passive and semi-passive UHF RFID systems. *IEEE Trans. Microw. Theory Tech.* 60, 1175–1182.

[11] Stewart, J. T., and Reynolds, M. S. (2012). "A 96 Mbit/sec, 15.5 pJ/bit 16-QAM modulator for UHF backscatter communication," in *Proceedings of the 2012 IEEE International Conference on RFID (RFID)*, (Warsaw: IEEE), 185–190.

[12] Winkler, M., Faseth, T., Arthaber, H., and Magerl, G. (2010). "An UHF RFID tag emulator for precise emulation of the physical layer," in *Proceedings of the Microwave Conference (EuMC 2010)*, European, Paris, 273–276.

[13] Pozar, D. M. (2012). *Microwave Engineering*. Hoboken: Wiley.

[14] John, K., and Tentzeris, M. M. (2016). "Pulse shaping for backscatter radio," in *Proceedings of the 2016 IEEE MTT-S International Microwave Symposium (IMS)*, (Honololu, HI: IEEE), 1–4.

[15] Matthew, S. T., Valenta, C. R., Koo, G. A., Marshall, B. R., and Durgin, G. D. (2012). "Multi-antenna techniques for enabling passive RFID tags and sensors at microwave frequencies," in *Proceedings of the 2012 IEEE International Conference on RFID (RFID)*, (Warsaw: IEEE), 1–7.

[16] Correia, R., Boaventura, A., and Borges Carvalho, N. (2017). "Quadrature amplitude backscatter modulator for passive wireless sensors in IoT applications," in *Proceedings of the IEEE Transactions on Microwave Theory and Techniques*, (Nanjing: IEEE), 1–8.

9

Beam-type Wireless Power Transfer and Solar Power Satellite

Naoki Shinohara

Research Institute for Sustainable Humanosphere, Kyoto University, Japan

Abstract

A wireless power transfer (WPT) through radio waves can be utilized not only for power diffusion (e.g., in a wireless communication system) but also for a "beam-type" WPT that might replace a wired power transfer system. In beam-type WPTs, microwaves are often used to focus the wireless power onto a single target with a high efficiency that is close to 100%. WPTs that utilize microwaves are referred to as microwave power transfers (MPTs), and MPTs can be applied for 1) long-distance beam-type WPTs to a fixed target, 2) mid- and short-distance beam-type WPTs to a fixed target, 3) beam-type WPTs to a moving target, and 4) Solar Power Satellites. Previous experiments and future systems with beam-type WPTs are described in this chapter.

9.1 Introduction

Electricity can be transmitted wirelessly through radio waves instead of a power line. It is possible to concentrate the wireless power into a single receiver with a high efficiency at a long distance. However, based on Maxwell's equation, larger antennas or higher frequencies than what are used with wireless power transfers (WPTs) are required to concentrate the wireless power. Microwaves (approximately 1–30 GHz) are usually utilized for long-distance WPTs to reduce the size of the antennas and to increase beam efficiencies. A WPT system that uses microwaves is referred

231

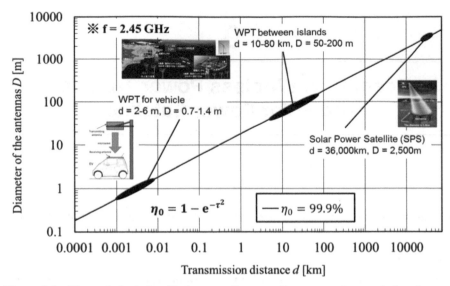

Figure 9.1   Theoretical relationship between diameter of antenna and transmission distance at 2.45 GHz with 99.9% beam efficiency.

to as a Microwave Power Transfer (MPT). Figure 9.1 indicates the relationship between the diameter of the antennas and the distance between a transmitting antenna and a receiving antenna when the beam efficiency is 99.9%. For example, the diameter of a transmitting antenna and a receiving antenna at distance of 10 km (80 km) is theoretically 50 m (200 m) with a 99.9% beam efficiency at 2.45 GHz, which is larger than desired for MPT. However, a 2.45 GHz WPT system with 99.9% efficiency can be realized theoretically even for Solar Power Satellites (SPSs), which are located 36,000 km above the surface of the Earth. We should also consider the efficiencies of the dc–RF conversion and the RF–dc conversion at the antennas to evaluate the total MPT efficiency. In 1975, W.C. Brown carried out WPT experiments in his laboratory at 2.45 GHz with a magnetron and a horn antenna (Figure 9.2) [1]. He achieved an efficiency of 54% at an output of 495 WDC in the experiment, which is the world record for a completely beam-type WPT system. Additionally, the beam direction can be controlled electrically with a phased array antenna and a retrodirective target detecting system. The wireless power can be supplied to a driving vehicle or flying drone. The history of WPTs via radio waves began with beam-type WPT applications in 1960s. Recently there have been many worldwide investigations into beam-type WPTs.

Figure 9.2 WPT laboratory experiment at 2.45 GHz by W.C. Brown in 1975 [1].

9.2 Long-Distance Beam-type WPT to Fixed Target

An image of a beam-type WPT (i.e., a MPT) is shown in Figure 9.3. Possible MPT applications of interest include MPT to an island, WPT to the top of a mountain, and MPT to an isolated place during an emergency instead of a power line. We must consider the initial cost and the running cost of the WPT system because we have other technologies to transmit electricity to the isolated place instead of the MPT system (e.g., via a power line or by carrying a battery).

Figure 9.3 Image of beam-type WPT to fixed target.

Figure 9.4   Beam-type WPT field experiment in 1994–95 in Japan.

There have been a few field experiments with beam-type WPTs. As shown in Section 9.1, W. C. Brown, Richard Dickinson, and their team succeeded in 1975 with the largest beam-type WPT demonstration up to that time at the Venus Site of the JPL Goldstone Facility (USA). The distance between the transmitting parabolic antennas whose diameter was 26 m, and the rectenna array, whose size was 3.4 m × 7.2 m, was 1 mile. The transmitted 2.388-GHz microwave signal was 450 kW from the klystron, and the rectified dc power achieved was 30 kW dc with an 82.5% rectifying efficiency. The estimated beam efficiency was approximately 8%.

In 1994–95 in Japan, Kyoto University and Kobe University carried out a beam-type WPT experiment at 2.45 GHz (Figure 9.4) supported by a Japanese power company. The researchers considered a beam-type WPT to an isolated place (e.g., the top of a mountain or an island). In the experiment, a 3 mf parabolic antenna and a 5 kW industrial magnetron were used as the transmitter and a rectenna array with 2,304 elements on a 3.2 m × 3.6 m grid served as the receiver [2]. The transmission distance was approximately 42 m, and the estimated beam efficiency was over 74%.

In 2008, Kobe University, a Japanese team, and a US team succeeded in a 148 km-distance MPT experiment with phased array on the crest of Haleakala on the island of Maui (Figure 9.5). This experiment was supported by Discovery Channel TV. The beam efficiency was very small, which was expected from the theoretical calculations, but the microwave power reached the receiving point.

In Europe, some unique technologies are presently being developed. Beam-type WPT field experiments on Réunion Island were planned in the late 1990s (Figure 9.6) [3, 4] in which a large rectenna array was developed. However, the project has not yet been carried out.

Figure 9.5 148 km distance MPT demonstration supported by Discovery Channel in 2008.

Figure 9.6 Grand Bassin, Réunion, France and their Prototype Rectenna [4].

As explained in Section 9.3, higher frequencies are suitable for beam-type WPTs with high beam efficiencies. In the US and in Japan, experiments on optical wireless power transfer via laser beams were carried out in the 2000 s. These experiments utilized a laser with a wavelength of approximately 1 mm to transmit wireless power of a few hundred watts. The antenna sizes and beam efficiencies of optical WPT systems are better than those of MPT systems, but there are problems with optical WPT systems that include accurate beam forming and the conversion efficiency between the dc and the laser.

In long-distance beam-type WPT systems, we must ensure the safety of the microwaves because it is possible that humans or animals will enter the microwave beam. Additionally, a study of the impact of the WPT system on conventional wireless communication systems is also important. However, we can divide the space of the microwave beam into "inside" and "outside" to reduce the influence of the microwave power on conventional wireless communication systems, which will be outside of the microwave beam. The safety problem of the radio waves and the study of the impact on the conventional wireless communication systems are the most important for realizing an MPT system in the near future.

9.3 Mid- and Short-Distance Beam-type WPT to Fixed Target

MPTs can also be utilized in short-distance applications, in which the size of the antennas can be reduced. In a short-distance MPT system, microwave power can be increased to the kW range because there will be no humans or animals between the transmitting and receiving antennas. Compared with the long-distance MPT system, the effective area of the microwave power in the short-distance MPT system is smaller, so an impact study is easier than in the long-distance MPT system.

One short-distance MPT application is the wireless charging of an electric vehicle (EV). Beginning in 2000, Kyoto University proposed and developed an MPT technique for an EV with Nissan Motors Co. (Figure 9.7). Transmitting antennas were on a road and rectennas were on the bottom of the EV. The distance between the transmitting and the receiving antennas was approximately 12.5 cm, which is 1 λ of 2.45 GHz. The battery on the EV can be charged using only microwave transmission with a theoretical beam efficiency of 83.7% and an experimental beam efficiency of 76.0% [5, 6]. This efficiency is high enough to economically transmit wireless power with microwaves. From 2006 to 2008, Mitsubishi Heavy Industries (MHI), Ltd. in Japan conducted an MPT research and development project for EVs with Mitsubishi Motors Co., Fuji Heavy Industries, Ltd, Daihatsu Motor Co., Ltd., and Kyoto University [7]. To reduce the power loss, they used (1) 6.6 kV to directly drive the 2.45-GHz magnetrons as microwave transmitters, (2) a blocking wall around which microwaves pass between the transmitting antennas and receivers, and (2) a heat recycling system. The total efficiency, including the heat recycling, was approximately 38% with an output power of 1 kW at a distance of 12.5 cm. The prototype released in 2009 is shown in Figure 9.8.

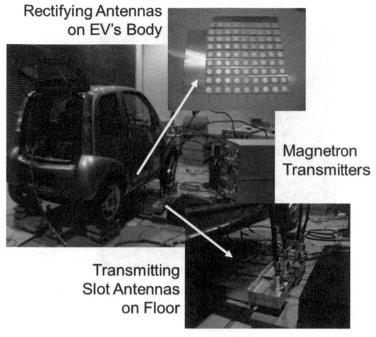

Figure 9.7   Wireless charging system of EV through microwave by Kyoto University and Nisson Motors Co. in 2004.

Figure 9.8   Wireless charging experiment with microwaves by Mitsubishi Heavy Industries's group in 2009.

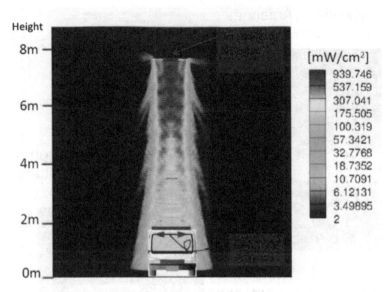

Figure 9.9   Proposed mid-distance wireless charging for EV and the FDTD simulation of the microwave beam [7].

In 2012, Volvo Technologies (Japan group), Nihon Dengyo Kosaku Ltd, and Kyoto University began to develop a new MPT system for an electric track. The former system led to mutual coupling problems between the transmitting and receiving antennas because the MPT distance was too short; therefore, the new MPT system was changed from a road-to-body to a top-to-roof configuration (Figure 9.9) [7, 8] to take advantage of the MPT as long-distance WPT. The distance between the transmitting antennas and the receiving antennas on the roof-top of the EV was 2–6 m, depending upon the type of the EV used. To maintain a high efficiency over the varying distances, a phased array system that could create a flat beam on the receiving antennas was proposed. On July 6th, 2012, Volvo Technologies, Japan and Nihon Dengyo Kosaku, Ltd released a 10-kW rectenna array with an efficiency of 84% operating at a frequency of 2.45 GHz for a mid-distance WPT system (Figure 9.10) [9]. The received microwave power density was more than 3.2 kW/m^2 at a distance of approximately 4 m from the transmitter.

In 2015, Kyoto University and the MHI group developed a wireless charging system for a motor bicycle at 2.45 GHz with a 100-W microwave (Figure 9.11). The rectenna was placed in front of a basket and could provide approximately 20–30 WDC from the received microwave. The dc power was

Figure 9.10 Photograph of the 10 kW Rectenna operating at 2.45 GHz for Wireless charging of EVs [8].

Figure 9.11 Wireless charger of motor bicycle at 2.45 GHz with a 100 W microwave.

provided directly to charge a battery in the motor bicycle. This charging system is very convenient because a user is not required actively to charge the battery. Starting in 2017, MPT systems will be investigated for daily use in the town hall in the south of Kyoto prefecture. For the experiment, the Ministry of Internal Affairs and Communications, which controls all radio wave applications in Japan, will allow a 'special zone for the MPT' in the town hall in the south of Kyoto prefecture.

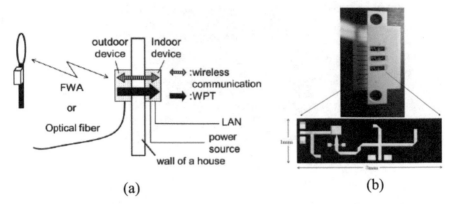

Figure 9.12   (a) Concept of a microwave-driven FWA system [10], (b) Developed 24 GHz MMIC rectenna [11].

The other proposed short-distance beam-type WPT is a Fixed Wireless Access (FWA) as proposed by NTT Corp. and Kyoto University. Figure 9.12(a) shows an image of the proposed system [10]. The outdoor device communicates with the Internet by the FWA or an optical fiber. The inside device and the outside device communicate wirelessly with each other. The inside device transmits power to the outside device via microwaves. The outside device can operate with no battery. For the system, it is better that wireless information and power are carried by the same microwave carrier to reduce the size of the system. At first, a frequency of 24 GHz was selected and an MMIC rectenna with a class-F load output filter was developed [11]. The dimensions of the MMIC rectenna at 24 GHz were 1 mm × 3 mm on GaAs (Figure 9.12(b)), and the maximum RF–dc conversion efficiency was 47.9% for a 210-mW microwave input at 24 GHz with a 120-W load.

9.4 Beam-type WPT to Moving Target

For fixed targets, long-distance beam-type WPTs can be replaced by wired power transfer systems and short- and mid-distance beam-type WPTs can be replaced by wired power transfer systems with an inductive WPT system. Each replacement system has both pros and cons. However, we have no power transfer systems for moving targets except for beam-type WPT systems. In fact, power transfer to moving targets is the most suitable application of beam-type WPT systems. As explained in Section 9.1, the first beam-type WPT experiment was carried out to a flying drone by W. C. Brown in 1964

Figure 9.13 The Canadian SHARP Flight Experiment and the 1/8 Model Airplane in 1987 [13].

because Brown knew well that the most suitable application of a beam-type WPT is to a moving target system.

A Canadian group from the Communications Research Centre successfully conducted a fuel-free airplane flight experiment using MPT in 1987 called SHARP (Stationary High Altitude Relay Platform; Figure 9.13) [12, 13]. The researchers transmitted a 2.45-GHz, 10-kW microwave signal to a model airplane with a total length of 2.9 m and a wing span of 4.5 m that was flying more than 150 m above ground level.

In University of Alaska Fairbanks (USA), researchers revised Brown's MPT-drone experiment, developed a new magnetron amplifier [14, 15], and carried out the MPT-drone experiment in 1995. They demonstrated the MPT-drone in Kobe, Japan in 1995 (Figure 9.14). The drone was fixed by a guideline above a transmitting antenna and a magnetron. The propeller was rotated only by microwave power. Another American group from the University of Colorado and the National Higher French Institute of Aeronautics and Space developed a WPT system to fly a micro unmanned vehicle (micro-UAV) in 2015 (Figure 9.15) [16].

In Japan, researchers at Kyoto and Kobe Universities developed a new phased array for an MPT experiment to a flying airplane in 1992. The Japanese project was named MIcrowave Lifted Airplane eXperiment (MILAX). In the MILAX transmitter, 96 GaAs semi-conductor amplifiers and 4-bit digital phase shifters were connected to 288 antenna elements at 2.411 GHz, which corresponds to three antennas for each amplifier sub-array. The gain of the amplifier was 42 dB at a 0-dBm input.

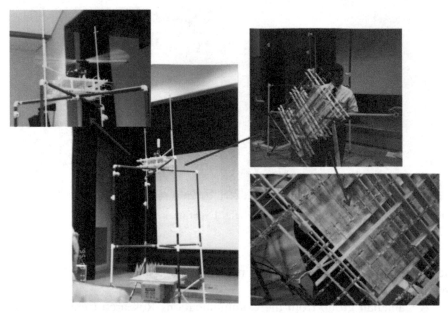

Figure 9.14   MPT-Drone Demonstration by University of Alaska Fairbanks in Kobe, Japan in 1995.

Figure 9.15   Experimental system of WPT to Micro-UAV by Univ. of Colorado.

The output power was approximately 42 dBm. The power added efficiency of the amplifier was approximately 40%. The diameter of the phased array was approximately 1.3 m. The beam width of the phased array was approximately

6 degrees. The total microwave power was 1.25 kW and consisted of a continuous wave with no modulation. The phased array was assembled on the roof of a car. The car drove under the fuel-free airplane for as much time as possible, and the microwave beam was directed toward the fuel-free model airplane using a computer and data from two charge-coupled device cameras that detected the position of the target. The rectenna array on the airplane's body is shown in Figure 9.16. There were 120 rectennas in all. The element spacing was 0.7 λ. The efficiency of the rectenna was approximately 61% at 1-W output dc power. The airplane flew freely 10 m above ground using only the microwave power that was delivered by the phased array. The maximum dc power obtained from the rectenna array was approximately 88 W, which was sufficient to fly the airplane (Figure 9.16). After the success of the MILAX experiment, the phased array was reused in the next MPT rocket experiment in 1993, which was named ISY-METS (Microwave Energy Transmission in Space in International Space Year) by Kyoto University, Kobe University,

Figure 9.16 MILAX microwave-driven airplane experiment with phased array in Japan in 1992.

Texas A&M University (USA), Communications Research Laboratory (CRL), National Institute of Information and Communications Technology, and the Institute of Space and Astronautical Science (ISAS).

Kobe University and CRL's group in Japan succeeded in an MPT field experiment involving a flying airship in 1995 called the ETHER (Energy Transmission toward High-altitude long endurance airship ExpeRiment) project. This research group transmitted 2.45-GHz, 10-kW microwaves to a flying airship 35–45 m above ground level (Figure 9.17). In these experiments, unlike in the MILAX project, researchers adopted a parabolic antenna MPT system using a microwave tube. Only a phased array system was used in the MILAX project, so ETHER was the first MPT field experiment to employ such an antenna system.

Ritsumeikan University's group in Japan carried out a WPT demonstration to a flying drone in 2015 (Figure 9.18) [17] using a 430-MHz frequency band and approximately 30-W radio wave power to supply power to the drone. The weight of the drone model was 25 g and it required 2 WDC to fly. At the time of this writing, the drone could fly above a transmitting antenna at distance of approximately 10 cm, but the system is under revision.

Figure 9.17   ETHER microwave-driven airship experiment with parabolic antenna in Japan in 1995.

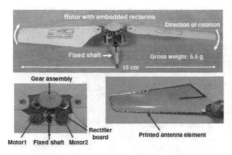

Figure 9.18 Experimental system of WPT to Flying Drone [17] and its demonstration with Revised Drone in Japan (March 2016, in Japan).

9.5 Solar Power Satellite (SPS)

As explained in Section 9.1, the research and development of MPT systems after the 1960s was prompted by the concept of SPSs. An SPS is a very large satellite designed as an electric power plant that orbits in a Geostationary Earth Orbit (GEO) 36,000 km above the Earth's surface. A typical image of an SPS is shown in Figure 9.19. The SPS system is composed of a space segment and a ground site that receives the power. The space segment consists of a solar energy collector to convert the solar energy into dc electricity, a dc-to-microwave converter, and a large antenna array to beam down the microwave power to the ground. The solar collector can be either photovoltaic cells or a solar thermal turbine. The dc-to-microwave converter can consist of a microwave tube system and/or a semi-conductor system. The third segment is a large-phased array. The power beam must be controlled accurately to less than 0.0005 degrees to the ground segment (i.e., the rectenna array).

An SPS system cannot be realized without MPT. MPT systems always require a "killer application" (i.e., some extremely useful functionality that can only work with that system) for commercial applications because the MPT system size is theoretically larger than what we envision. Only SPS can serve as the killer application for MPT systems. The sizes of the transmitting and receiving antennas are theoretically over 2 kmϕ with a 90% beam efficiency at 5.8 GHz, but 1 GW of electricity would be obtained on the ground from a few square kilometers of solar cells in the SPS. If SPS is designed as a satellite, it would weigh over 10,000 tons, which is extremely large. However, we could acquire approximately 7–10 times more electricity in a more stable fashion with SPS than from solar cells on the ground. This is due to the fact that there is no night in space above 36,000 km and we

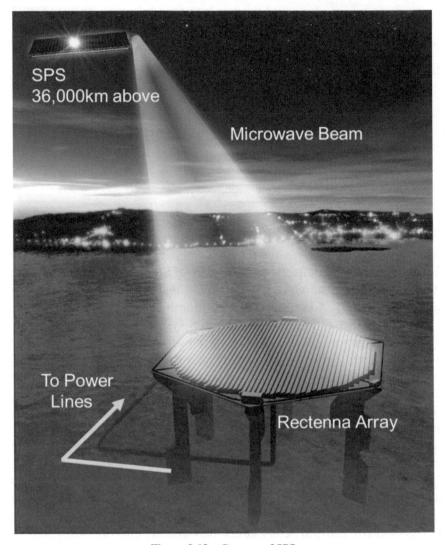

Figure 9.19   Concept of SPS.

can receive microwave power even on rainy days because microwaves can propagate without loss through rain, cloud, atmosphere, and ionosphere from space to ground.

After proposal of the first SPS by P. E. Glaser in 1968, there have been several SPS systems designed worldwide. Figure 9.20 indicates various the SPS designs.

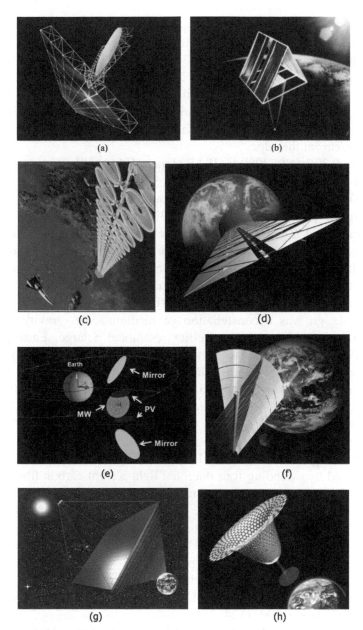

Figure 9.20 Various SPS designs (a) Reference System by NASA/DOE (US, 1978), (b) SPS2000 by ISAS (Japan, 1992), (c) Sun Tower by NASA (US, 1997), (d) Solar Sail (EU, 1990s), (e) JAXA2004 Model by JAXA (Japan, 2004), (f) Laser SPS by JAXA (Japan, 2004), (g) USEF2006 Model by USEF/METI (Japan, 2006), (h) SPS-alpha by NASA (US, 2012).

- Figure 9.20(a) Reference System by National Aeronautics and Space Administration (NASA)/Department of Energy (DOE) (US, 1978) [18] Separated solar cells (50 km^2, 13%, 9 GWDC output) and microwave antennas (1 kmϕ, 6.72 GWMW, 2.45 GHz) are connected via a rotary joint at 36,000 km altitude. On the ground, a 13 km $\times$ 10 km rectenna array receives the microwaves and converts them to dc (5 GW). The power density is 23 mW/cm^2 at the center of the rectenna array, but just 1 mW/cm^2 at the edge.

- Figure 9.20(b) SPS2000 by ISAS (Japan, 1993) [19]
  This was a middle-size experimental SPS in equatorial low Earth orbit (1,100 km altitude). Solar cells would be on a shape similar to a triangular prism with a length of 303 m and sides of 336 m. The frequency of 2.45 GHz was assigned to transmit the power from a transmitting antenna whose dimensions are 132 m $\times$ 132 m to the Earth. SPS2000 can exclusively serve the equatorial zone, particularly benefiting geographically isolated lands in developing nations.

- Figure 9.20(c) Sun Tower by NASA (US, 1997) [20]
  This design was a constellation of medium-scale, gravity gradient-stabilized satellites. Each satellite resembled a large, Earth-pointing sunflower in which the face of the flower is the transmitter array and the "leaves" on the stalk were solar collectors. The concept was assumed to transmit at 5.8 GHz from an initial operational orbit of 1000 km and be sun-synchronous at a transmitted microwave power level of about 200 MW.

- Figure 9.20(d) Sail Tower (EU, 1990s) [21–23]
  Each sail has dimensions of 150 m $\times$ 150 m and was automatically deployed by extending four diagonal light-weight carbon fiber (CFRP) booms that were initially rolled up on a central hub. Researchers demonstrated the functionality of the CFRP booms. The power generated within the sail modules would be transmitted through the central tether to the antenna, where 2.45 GHz microwaves would be generated in mass-produced inexpensive magnetrons.

- Figure 9.20(e) JAXA2004 Model by JAXA (Japan Aerospace Exploration Agency, formerly NASDA, the National Space Development Agency) (Japan, 2004) [24, 25]
  This design was a 5.8-GHz, 1-GW SPS, and it was called the "Formation Flying Model." The design is based on the formation-flight of a rotating mirror system and an integrated panel composed of a solar cell surface on one side and a phased microwave array antenna on the other side at

36,000 km altitude. The buoyancy can be used to fly the primary mirrors independently. Formation flying mirrors are used to eliminate the need for rotary joints. The size of the primary mirrors was 2.5 km × 3.5 km (×2 panels), and their weight was 1,000 tons each. The integrated main panel would be composed of a solar cell of size 1.2–2 kmφ, a microwave transmitter and antennas of 1.8–2.5 kmφ, and two secondary mirrors, whose total weight was approximately 8,000 tons.

- Figure 9.20(f) Laser SPS by JAXA (Japan, 2004)
  This design is not a microwave SPS but rather a laser SPS. The size of the design is 400 m × 200 m × 12 km with 100 modules, and its weight is 5,000 tons. The wavelength of the laser was 1.064 mm from a Nd/Cr:YAG Ceramics laser, and the laser would be directly converted from sunlight with a 19.1% efficiency. The power of the transmitting laser was in the 10 MW–100 MW class.
- Figure 9.20(g) USEF2004 Model by USEF/METI (Japan, 2004) [26, 27]
  This design is called Tethered-SPS. The attitude is automatically stabilized by the gravity gradient force in the tether configuration without any active attitude control. The design is composed of a 2.0 km × 1.9 km power generation/transmission panel suspended by multi-wires deployed from a bus system that is located over 10 km from the power panel. Because of the power generation/transmission panel and the gravity gradient force in the tether configuration, the generated power from the sun cannot be stable and therefore changes with time. The design is capable of 1.2 GW power transmissions and 0.75 GW average power receptions on the ground. The weights of the panel and the bus system were 18,000 tons and 2,000 tons, respectively.
- Figure 9.20(h) SPS-alpha by NASA (US, 2012) [28, 29]
  This design involves three major functional elements: (1) a large primary array that points toward Earth; (2) a very large sunlight-intercepting reflector system consisting of a large number of reflectors that act as individually pointing "heliostats" mounted on a non-moving structure; and (3) a truss structure that connects the first two components.

Table 9.1 shows some typical parameters of the transmitting antennas of the SPS. An amplitude taper on the transmitting antenna is adopted in order to increase the beam collection efficiency and to decrease the side-lobe levels in almost all SPS designs. A typical 10 dB Gaussian amplitude taper is used, in which the power density in the center of the transmitting antenna is ten times larger than the density on the edge of the transmitting antenna.

Table 9.1   Typical parameters of the transmitting antenna in an SPS

Model	Reference Model	JAXA Model
Frequency	2.45 GHz	5.8 GHz
Diameter of transmitting antenna	1 kmϕ	1.93 kmϕ
Amplitude taper	10 dB Gaussian	10 dB Gaussian
Output power (beamed to Earth)	6.72 GW	1.3 GW
Maximum power density at center	2.2 W/cm^2	114 mW/cm^2
Minimum power density at edge	0.22 W/cm^2	11.4 mW/cm^2
Antenna spacing	0.75 λ	0.75 λ
Power per antenna	Max. 185 W	Max. 1.7 W
(Number of elements)	(97 millions)	(1,950 millions)
Rectenna Diameter	10 km × 13.2 km	2.45 kmϕ
Maximum Power Density	23 mW/cm^2	100 mW/cm^2
Collection Efficiency	89%	87%

Details of the microwave beam in the JAXA (Japan Aerospace eXploration Agency) model, which was designed in 2004, are shown in Figure 9.21. The microwave power must remain below 1 mW/cm^2 outside of the rectenna. Compared to the NASA/DOE model that was designed in 1978, the microwave power density on the ground for the JAXA model is greater because Japan is a small country and the ground system must therefore be smaller. For the parameters shown, the MPT system in the SPS is a radiative near-field system even at 36,000 km altitude because the antenna gain is extremely large.

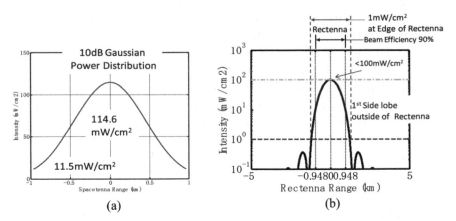

Figure 9.21   JAXA2004 Model (a) Power distribution of transmitting antenna whose size is 1.93 km and whose microwave power is 1.3 GW (b) Power distribution of rectenna on ground whose size is 2.45 km.

References

[1] Brown, W. C. (1973). "Adapting microwave techniques to help solve future energy problems", *1973 G- MTT International Microwave Symposium Digest of Technical Papers*, Seattle, WA, 189–191.

[2] Shinohara, N., and Matsumoto, H. (1998). Experimental study of large rectenna array for microwave energy transmission. *IEEE-Trans. MTT*, 46, 261–268.

[3] Celeste, A., Luk, J-D. L. S., Chabriat, J. P., and Pignolet, G. (1997). "The grand-bassin case study: technical aspects", *Proceedings of the SPS'97*, Montreal, 255–258.

[4] Celeste, A., Jeanty, P., and Pignolet, G. (2004). Case study in Reunion island. *Acta Astronaut.* 54, 253–258.

[5] Shinohara, N., and Matsumoto, H. (2004). Wireless charging system by microwave power transmission for electric motor vehicles (in Japanese). *IEICE Trans. C*. J87-C, 433–443.

[6] Shinohara, N. (2011). Beam efficiency of wireless power transmission via radio waves from short range to long range. *J. Korean Inst. Electro. Eng. Sci.* 10, 224–230.

[7] Shinohara, N. (2013). "Wireless power transmission progress for electric vehicle in Japan," *Proceedings of the 2013 IEEE Radio & Wireless Symposium (RWS)*, Tokyo, 109–111.

[8] Shinohara, N., and Kubo, Y. (2013). Suppression of unexpected radiation from microwave power transmission system toward electric vehicle. *Proceedings of the 2013 Asia-Pasific Radio Science Conference (AP-RASC)*, Tokyo, E3–4.

[9] Furukawa, M., Minegishi, T., Ogawa, T., Sato, Y., Wang, P., Tonomura, H. (2013). "Wireless power transmission to 10 kW output 2.4 GHz-band rectenna array for electric trucks application (in Japanese)", *IEICE Tech. Report, WPT2012-47*, 36–39.

[10] Hatano, K., Shinohara, N., Mitani, T., Seki, T., and Kawashima, M., (2012). "Development of improved 24 GHz-Band Class-F load rectennas," in *Proceedings of the 2012 IEEE MTT-S International Microwave Workshop Series on Innovative Wireless Power Transmission: Technologies, Systems, and Applications (IMWS-IWPT2012)*, Tokyo, 163–166.

[11] Hatano, K., Shinohara, N., Seki, T., and Kawashima, M. (2013). "Development of MMIC rectenna at 24 GHz," in *Proceedings of 2013 IEEE Radio & Wireless Symposium (RWS)*, Austin, TX, 199–201.

[12] Schlesak, J. J., Alden, A., and Ohno, T. (1988). "A microwave powered high altitude platform," in *Proceedings of IEEE MTT-S International Symposium Digest*, Baltimore, MD, 283–286.

[13] SHARP. Available at: http://www.friendsofcrc.ca/Projects/SHARP/sharp.html

[14] Hatfield, M. C., Hawkins, J. G., and Brown, W. C. (1998). "Use of a magnetron as a high-gain, phase-locked amplifier in an electrically-steerable phased array for wireless power transmission," in *Proceedings of 1998 MTT-S International Microwave Symposium*, Baltimore, MD, 1157–1160.

[15] Hatfield, M. C., and Hawkins, J. G. (1999). "Design of an electronically-steerable phased array for wireless power transmission using a magnetron directional amplifier," in *Proceedings of 1999 MTT-S International Microwave Symposium*, Baltimore, MD, 341–344.

[16] Dunbar, S., Wenzl, F., Hack, C., Hafeza, R., Esfeer, H., Defay, F., et al. (2015). "Wireless far-field charging of a micro-UAV," in *Proceedings of IEEE Wireless Power Transfer Conference (WPTc)*, Boulder, CO, T1.2.

[17] Nishikawa, H., Kiani, Y., Furukoshi, T., Yamaguchi, H., Tanaka, A., and Douseki, T. (2015). "UHF power transmission system for multiple small self-rotating targets and verification with batteryless quadcopter having rotors with embedded rectenna," in *Proceedings of IEEE Wireless Power Transfer Conference (WPTc)*, Boulder, CO, T1.1.

[18] DOE and NASA report (1978). *Satellite Power System; Concept Development and Evaluation Program*. Washington, DC: NASA.

[19] SPS 2000 Task Team (1993). *SPS 2000 Project Concept – A Strawman SPS System*. Japan: S2-T1-X, Preliminary.

[20] Mankins, J. C. (1997). A fresh look at space solar power: new architectures, concepts and technologies. *Acta Astronaut.* 41, 347–359.

[21] Seboldt, W., Klimke, M., Leipold, M., and Hanowski, N. (2001). European sail tower SPS concept. *Acta Astronaut.* 48, 785–792.

[22] Leipold, M., Eiden, M., Garner, C. E., Herbeck, L., Kassing, D., Niederstadt, T., et al. (2003). Solar sail technology development and demonstration. *Acta Astronaut.* 52, 317–326.

[23] A Study for ESA Advanced Concept Team (2005). *Earth & Space-Based Power Generation Systems – A Comparison Study*. Paris: ESA.

[24] Mori, M., Nagayama, H., Saito, Y., and Matsumoto, H. (2004). Summary of studies on space solar power systems of the national space development agency of Japan. *Acta Astronaut.* 54, 337–345.

[25] Oda, M. (2004). "Realization of the solar power satellite using the formation flying solar reflector," in *Proceedings of the NASA Formation Flying Symposium*, Washington, DC.

[26] Sasaki, S., Tanaka, K., Kawasaki, S., Shinohara, N., Higuchi, K., Okuizumi, N., et al. (2004). Conceptual study of SPS demonstration experiment. *Radio Sci. Bull.* 310, 9–14.

[27] Fuse, Y., Saito, T., Mihara, S., Ijichi, K., Namura, K., Honma, Y., et al. (2011). "Outline and progress of the Japanese microwave energy transmission program for SSPS," in *Proceedings of the 2011 IEEE MTT-S International Microwave Workshop Series on Innovative Wireless Power Transmission: Technologies, Systems, and Applications (IMWS-IWPT2011)*, Kyoto, 47–50.

[28] Mankins, J. C. (2009). New detections for space solar power. *Acta Astronaut.* 65, 146–156.

[29] SPS-ALPHA (2012). *The First Practical Solar Power Satellite via Arbitrarily Large Phased Array (A 2011–2012 NASA NIAC Phase 1 Project)*. Available at: https://www.nasa.gov/pdf/716070main_Mankins_2011_PhI_SPS_Alpha.pdf

PART III

Coexistence of WPT

10

Human Safety on Electromagnetic Fields – The International Health Assessment

Junji Miyakoshi

Kyoto University, Japan

10.1 Introduction

Since the late 20th century, the environment has become increasingly flooded with electromagnetic fields, due to the rapid worldwide proliferation of electromagnetic field sources such as cellphones and base stations, high-voltage lines, electric appliances, medical facility equipment, and so on. In the near future, practical development of wireless power transfer through electromagnetic fields is expected to add substantially to this trend. The continuing trend in society and daily life will be an inexorably rising electromagnetic environment of static magnetic fields, low-, intermediate-, and radiofrequency fields, millimeter waves, terahertz waves, and other electromagnetic fields. Like ionizing radiation, these electromagnetic fields are invisible and therefore create unease in many people about the possible health effects.

The history of research devoted to the effects of low-frequency and radiofrequency fields on health is very short and shallow compared to research on ionizing radiation. In this report, the current state of domestic and foreign research on the biological effects of non-ionizing electromagnetic fields and assessments performed by the World Health Organization (WHO), the International Agency for Research on Cancer (IARC), and other international organizations are summarized. The main focus is the pursuit of an accurate scientific understanding of the biological effects of electromagnetic fields, which remain largely unknown. In this section, the author discussed electromagnetic fields as an environmental factor in everyday life and in particular summarized the international assessments of the health effects of

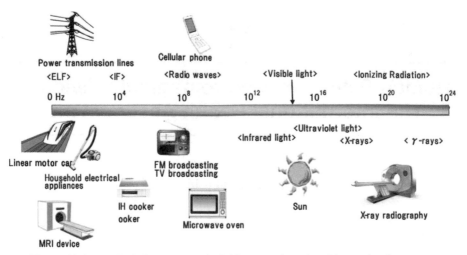

Figure 10.1   Typical electromagnetic field sources in order of increasing frequency.

extremely low-frequency and radiofrequency electromagnetic fields. Those of static magnetic fields and low and high frequencies (including utility frequencies) have been described in other published reports [1–3].

10.2 Historical Background on Electromagnetic Fields and Health

The historical background on electromagnetic fields and health begins with a report of high leukemia incidence in children living near high-voltage lines by American epidemiologists in 1979 [4], and since the early 1990s has been marked by highly active international studies and discussions on the health effects of non-ionizing electromagnetic fields such as a static magnetic field, low-frequency and radiofrequency electromagnetic fields. The present and future possible sources of exposure to electromagnetic fields, however, include the strong static magnetic fields produced during magnetic resonance imaging, the extremely-low frequency electromagnetic fields of commercial frequency regions, the radiofrequency regions utilized by cellphones, the electromagnetic fields of wireless power transfer, which is nearing practical development, the intermediate-frequency band electromagnetic fields of induction-heating ranges, and the millimeter-wave imaging equipment used in airports and other facilities Figure 10.1 shows typical non-ionizing and ionizing electromagnetic field sources, in order of increasing frequency.

Following the 1979 epidemiological report and into the 1990s, biological research with animals and cells was particularly active in Europe, the U.S., and Japan, and numerous epidemiological studies on extremely-low frequency electromagnetic fields (particularly on magnetic fields) from high-voltage lines were conducted. Since the rapid rise of cellphones in the late 1990s, international studies and discussions on radiofrequency have also become highly active.

The course of research on electromagnetic fields and assessments of their effects on health to the present is essentially as follows.

10.3 Studies Relating to Assessment of Electromagnetic Field Effects on Health

10.3.1 Overview

The known biological effects of non-ionizing electromagnetic fields are generally divided into two categories: those below and those above a frequency of approximately 100 kHz. It is known that the low-frequency band below 100 kHz exerts a "stimulating action" and the frequency band above 100 kHz exerts a "thermal action". Cell and animal-level biological studies on extremely-low frequency electromagnetic fields have generally indicated that the effects of electromagnetic fields are absent at the levels encountered in the environment (generally 1 μT or less) and begin to appear at several tens of thousand times that level (several mT). Most studies on biological effects of electromagnetic fields focus on the low magnetic flux densities common in residential environments and thus on very low levels of exposure, and may therefore be reasonably expected to show no substantial effect on cells or animals.

Exposure to radiofrequency electromagnetic fields occurs in the clinical use of strong microwaves for their thermal effect in cancer, rheumatic, and neuralgic thermotherapy. However, there has been little research on radiofrequency electromagnetic fields at the environmental level and in many respects this topic remains uncertain. As described above, the worldwide proliferation of cellphones since the late 1990s has been accompanied from the beginning by unease concerning the possibility of inducing brain tumors and other adverse effects through the emitted radiofrequency electromagnetic fields, particular in view of their use in close proximity to the brain. Debate has also heightened on the possibility of "non-thermal actions" and particularly their effects on children.

Table 10.1   Main methodologies of research on biological effects of electromagnetic fields

Epidemiological study	Investigation of electromagnetic fields and health (mainly carcinogenesis) for humans
Animal study	Experiments on biological assessment for electromagnetic fields using animals such as mouse and rat
Cellular study	Experiments on cellular and genetic assessment for electromagnetic fields using cells derived from humans and animals

The main methods (Table 10.1) used in the study of electromagnetic field-biological effects generally involve (1) human epidemiology or volunteers, (2) animal experiments, or (3) cellular experiments.

Basic differences between humans, animals, and cells as the objects of research preclude pronouncement of the superiority of one over the others, but the weighting of assessments on the effects on humans generally increases in the order of human epidemiology, animal experiment, and cellular experiment. The accuracy and reproducibility of assessment methods, on the other hand, generally increase in the order of cellular experiment, animal experiment, and epidemiology, and the length of research also increases in this order (Figure 10.2). It may also be noted that the weighting of cellular research has increased substantially with its rising use of deoxyribonucleic

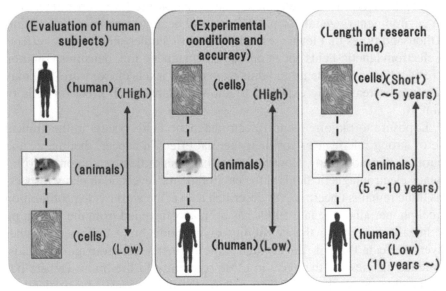

Figure 10.2   Categories of epidemiological and biological (animal and cellular) research.

Table 10.2 Main indicators in studies on the biological effects of electromagnetic fields

Research Category	Subject	Evaluation Criteria
In Vitro Study	Cells	Cell proliferation, DNA synthesis, chromosomal aberration, sister chromatid exchange, micronucleus formation, DNA strand breaks, gene expression, signal transduction, ion channels, mutation, transformation, immune responses, cell differentiation, cell cycle, apoptosis
In Vivo Study	Laboratory animals (rat, mouse, etc.)	Carcinogenesis (lymphoma, leukemia, brain tumors, mammary gland tumor, liver cancer), reproduction and development (implantation rate, fetal body weight, teratogenesis), abnormal behavior, neuroendocrinology mainly melatonin, immune function, blood brain barrier
Epidemiological Study	Human	Carcinogenesis and cancer death (brain tumor, childhood and adult leukemia, breast cancer, melanoma, lymphoma), reproductive ability, spontaneous abortion, Alzheimer disease
Influence on Human Body	Human	Psychological and physiological influences (fatigue, headache, anxiety, lack of sleep, brain waves, electrocardiogram, memory), neuroendocrinology mainly melatonin, immune function

acid (DNA) and genes since the completion of the Human Genome Project (HGP) more than ten years ago.

Table 10.2 is a summary of the main indicators that have been used in cells, animals, and individual human studies for assessment of electromagnetic field in biological effects. Much of the research has long been focused on assessment of electromagnetic field in carcinogenetic effects, but in recent years cellular research on the effects on immune response, stress response, apoptosis, and other functional manifestations has also attracted growing interest. In epidemiological research, in addition to cancer, Alzheimer's and other disorders have become a subject of study and the range of assessment indicators has accordingly broadened.

10.3.2 Epidemiological Studies

Epidemiological studies on humans tend to influence public opinion more strongly than cell and animal experiments, but human living environments vary, pure investigation of the factors under study is impracticable, and biased group selection and other factors (termed "selection bias" and "confounders") cannot be entirely excluded and may skew the results and the statistical assessments. As noted previously, the first report of a carcinogenic effect caused by extremely-low frequency electromagnetic fields was an epidemiological study performed in 1979, which led to rising international debate, the mounting of many other epidemiological studies on extremely-low frequency electromagnetic fields in Europe and the U.S. in the 1990s [5], and the first coordination of epidemiological studies in this field by the National Institute for Environmental Studies in Japan in the 2000s [6].

A nine-country pooled analysis of extremely-low frequency electromagnetic fields (more precisely defined as extremely-low frequency magnetic fields and childhood leukemia found no association between exposure of children living in environments of less than 0.4 μT (approximately 99.2% of households) to extremely-low frequency electromagnetic fields and the risk of leukemia, thus indicating "no effect". For those living in residential environments with extremely-low frequency electromagnetic fields at levels of 0.4 μT or higher (approximately 0.8% of children), the leukemia risk was almost twice as high, showing a statistical significance [7]. Similar trends were found in epidemiological research performed in Japan [6]. The results of epidemiological studies on other childhood cancers and on adult cancers indicate "no effect (no association found)" for extremely-low frequency electromagnetic fields. For those results of the epidemiologic studies that show an increase in childhood leukemia caused by extremely-low frequency electromagnetic fields, the biological action mechanisms remain unclear and, as noted earlier, the possibility of reduced accuracy due to selection bias and confounders cannot be completely ruled out.

The IARC and WHO assessments for extremely-low frequency electromagnetic fields were followed by the report of a new pooled analysis of extremely low-frequency magnetic field exposure and risk of childhood leukemia [8], in which the epidemiological studies on seven groups emphasized accurate measurement of electromagnetic field environments and the results of the above epidemiological study performed in Japan were included. The report concluded that there was no large difference from the nine-country pooled analysis results [7], and that no change should be made in the below-described WHO carcinogenic assessment or comprehensive assessment based on environmental health quality criteria.

Radiofrequencies related to cellphones have also been a subject of highly active international epidemiological research, including a major study named "The Interphone Study" and conducted by the IARC with participation by Japan, the United Kingdom, Sweden, and ten other countries (but not the U.S.), as a set of case-control studies focused on the risk of brain tumors. The IARC compiled the studies of all the participating countries and presented an overview of the final conclusions of this international joint cooperative research in a press release issued in May 2010 [9]. In summary, the results were: (1) the odds ratio for glioma and meningioma decreased somewhat in regular cellphone users; (2) no increase in the odds ratio was observed in long-term users (10 years or more); and (3) users with large cumulative call times (1,640 hours or more) showed a slight increase of 1.40 in the odds ratio for glioma (95% confidence interval: 1.03 to 1.89). The odds ratio here represents a statistical quantity for individuals with brain tumors, as the probability of being a cellphone user divided by the probability of being a non-user. In conclusion, the results indicate there is no increase in brain tumors (glioma and/or meningioma) among long-term cellphone users (10 years or more), and the report further states that accurate interpretation of a causal relation is difficult for the observed decreases in the odds ratio, increases in the odds ratio for long cumulative callers, increases in glioma at the dorsal fold on the cellphone-using side, and other such effects.

Most other epidemiological studies have similarly found no evidence of increased carcinogenicity. Exceptions include an epidemiological pooled study in Sweden showing a tripling of glioma occurrence in callers with cumulative calling times that exceeded 2,000 hours [10], and another in Japan indicating an increase in acoustic neuroma with calling times of 20 minutes or more per day [11]. No clear evidence has been found for association of occupational exposure to microwaves with brain tumors, leukemia, lymphoma, or other cancers, nor for association of radio-wave transmission from radio and television towers, base stations, or other facilities with carcinogenicity. Epidemiological research on childhood use of cellphones and carcinogenicity includes the Cefalo project with trinational participation (including Denmark) and the 15-country MobiKids project (including Japan). The Cefalo study has been completed, with no finding of a statistically significant effect [12, 13]. The Mobi-Kids study [14], which began in January 2014, has been renamed the EU GERoNiMO project [15]. It is a program of generalized electromagnetic field research using novel methods, with an integrated approach extending from research to risk assessment and support to risk management, and in the period of five years ending in 2018 is now working toward completion of a major new program that includes animal, cell, and molecular-level studies as well as the usual methods of epidemiological research, in a work package extends to risk management and communication.

The study plan is very broad, and includes the intermediate frequency band in its range of investigated frequencies.

10.3.3 Animal Experiments

Many animal experiments with mice and rats were performed in the 1990s for assessment of the biological effects of electromagnetic fields of extremely-low frequency electromagnetic fields. Most of these studies consisted almost entirely of investigation for carcinogenic effects, while some looked for associations with reproduction (fetal/embryonic development and teratogenicity), the nervous system (behavioral and sensory function), or immune function. One major question centered on whether exposure to extremely-low frequency electromagnetic fields might affect the carcinogenic process by transforming normal cells to cancerous cells (cancer cell initiation), promoting proliferation of the initiated cells (proliferation promotion), or both. The investigated extremely-low frequency electromagnetic fields flux densities ranged from several μT to 1 mT. The study results were nearly all negative (i.e., showing no carcinogenic effect), while a very small number indicated an increase in leukemia or breast tumors as a result of extremely-low frequency electromagnetic fields exposure [16]. Nearly all of the studies on processes other than carcinogenicity (i.e., reproduction, behavior, and immunity) also reported "no effect". In summary, the animal-experiment investigations performed to present have not found any clear effects of extremely-low frequency electromagnetic fields and have not provided sufficient evidence of their existence.

Radiofrequency electromagnetic fields have also been extensively investigated. In 1997, a study with transgenic mice was reported to have found an increase in leukemia due to radiofrequency electromagnetic exposure [17], and the 2000s was a time of increasingly active assessment of radiofrequency electromagnetic fields for carcinogenic effects, with animal experiments centering particularly in Europe, the U.S., and Japan. Among the studies reported to present, a few of which involved long-term (two-year) exposure and the use of animals particularly prone to carcinogenicity, nearly all found no effects attributable to radiofrequency electromagnetic fields [18]. Several studies on combined (chemical and radiofrequency electromagnetic fields) exposure have found increased carcinogenesis when radiofrequency electromagnetic fields are included in the combination [19–21].

An interim report [22] was given in June 2016 at the academic conference BioEM2016 (Ghent, Belgium) on the large-scale animal-experiment study conducted under the National Toxicological Program (NTP) of the National Institutes of Health (NIH) in the U.S. The report was essentially as follows:

1. Study conditions (excerpts)

   - Radiofrequency electromagnetic field exposures: CDMA, GSM, 1900 MHz (mice), 900 MHz (rats).
   - Animals: rats, mice (experiment point 1 group, 90 animals).
   - Exposure mode: whole-body, 1, 1.5, 3, 6 W/kg, 10-min ON/OFF (up to 9 hours), daily, 107 weeks.

2. Results (excerpts)

   - Lifespan: tendency for longer lifespan in exposed group than sham (control group).
   - Brain tumors: increased by exposure in male rats.
     (all exposures with GSM, 6 W/kg only with CDMA).
   - Cardiac schwannomas: SAR-dependent (up to 6 W/kg) increase in male rats with both GSM and CDMA (Cardiac schwannoma: primary heart neurilemmoma composed of schwann cells).
   - Female rats: No effect on brain or heart found at any exposure.
   - Cellular genotoxicity (not included in interim report): Specific absorption rate (SAR)-dependent increase in brain cells using comet assay (DNA strand break) test, but no effect in erythrocyte micronucleus formation (cell nucleus fragment separation) test.

3. Significance of interim report results

   - Cellphones are now in common use throughout the world.
   - The study content is of strong interest to the public and the media.
   - Its findings match those of some epidemiological studies.
   - Its findings support the IARC assessment results.

4. Scheduled course of NTP study

   - Completion of unanalyzed rat tissue assessment.
   - Performance of remaining assessments of mouse study results.
   - Completion of the pathological assessments in approximately 18 months.
   - Concurrent preparation for compilation of technical reports (TRs).

NIH studies are highly authoritative and carry great weight internationally, but this is an interim report and we await information on the further progress of this study and its completion.

10.3.4 Cellular Experiments

Research on the effects of electromagnetic fields on cells (and their genes and molecules) is highly active in many countries, and many papers have been issued. The space available here will not allow extensive details, but further information may be found in related literature [1–3]. Much of the research consists of investigations for associations of electromagnetic fields effects with carcinogenicity, cell genotoxicity (i.e., micronucleus formation, DNA damage, chromosome abnormality, and mutation), and functional changes in gene expression (e.g., oncogenes and expression of heat-shock and other stress proteins). Recently, significant relationship between carcinogenicity and micronuclear formation in cells was reported [23]. Figure 10.3 shows the formation of a micronucleus by chromosome separation, fragment formation, and cleavage, in which a chromosome fragment (DNA) that has separated from the nucleus emerges as a micronucleus in the bi-nucleated cell division phase.

In some early research with cellular experiments at the background electromagnetic field level of the environment (generally 1 μT or lower) the test results were positive for an effect but were later found lacking in reproducibility and considered to represent "no effect" or "effect too small for detection" [18].

Cellular experiments with radiofrequency electromagnetic fields emitted by cellphones and base stations have been performed in many studies in the EU, the U.S., Japan, South Korea, and other countries since 2000. The results have generally been negative for any cell genotoxicity involved in carcinogenicity at radiofrequency electromagnetic field levels with no heating effect. In a study focused on heat-shock proteins as a product of cellular

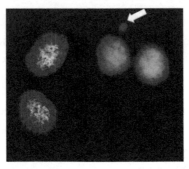

Figure 10.3   Representative formation of a micronucleus (arrow) in cultured bi-nucleated cell.

metabolic function, on the other hand, it was found that production of certain types of heat-shock protein (e.g., HSP-27) is increased as a non-thermal effect of radiofrequency electromagnetic fields [24]. Although this finding may be considered positive for a biological effect of cellphone and base-station radiofrequency electromagnetic fields and has been replicated in subsequent experiments, it has not been confirmed at many laboratories and some reports have been negative, and no clear scientific conclusion has emerged. In cellular studies testing for genotoxicity and non-genotoxicity (immune function, gene expression (RNA, protein), signal transduction, oxidative stress, apoptosis, proliferative capacity, and other tests) of radiofrequency electromagnetic fields, some positive results for genotoxicity have been reported, but no clear evidence has been found that they are attributable to a radiofrequency action mechanism under non-thermal conditions [18].

10.4 WHO and IARC Assessments, and Related Trends

WHO launched the International electromagnetic field project in 1996 [25], in the midst of the rising international debate on effects of electromagnetic fields on health that began in 1990, and since its launch the number of participating countries grown to 60. Organizationally, the project is affiliated with the WHO division for effects of ionizing radiation on health. It convenes symposia and workshops, and provides progress reports on biological effects assessment and proposals on problems to be addressed. Figure 10.4 shows the organization and role of WHO electromagnetic field for biological assessment.

The WHO-IARC convened a conference (Lyon, France) in 2001 on assessment of extremely-low frequency electromagnetic fields. It is essential to note that the IARC carcinogenicity assessment is qualitative and simply presents the strength of the evidence, without quantifying the degree of carcinogenic effects. Failure to keep this in mind has sometimes resulted in reports that invite public misunderstanding. In brief summary, the classifications were as follows.

1. Extremely-low frequency magnetic fields were classified as "Group 2B" (may be carcinogenic).
   The basis given for this Group 2B classification was study findings suggesting that extremely-low frequency magnetic fields increase childhood leukemia.

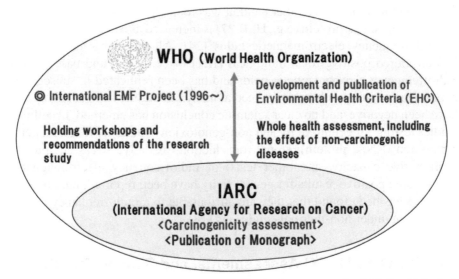

Figure 10.4   The organization and role of WHO electromagnetic field for biological assessment.

2. Static magnetic fields, static electric fields, and extremely-low frequency electric fields were classified as "Group 3" (non-classifiable in relation to human carcinogenicity). The basis given for this Group 3 classification was that data was insufficient for carcinogenicity assessment.

It may be noted that epidemiological study results strongly influenced the above Group 2B classification. Further information is available in the 80 IARC monographs [16].

In 2005 WHO convened a Task Group (including the author) to formulate the Environmental Health Criteria (EHC) for the assessment of biological effects from extremely-low frequency electromagnetic field, including those other than cancer. The final version was completed in approximately two years, and issued on the Internet in June 2007 [26] and in printed form in February 2008 [27]. It is a large, 519-page volume. All chapters are written in English and the content of Chapter 1, which is an important overview of the criteria, is also written in French, Russian, and Spanish at the end of that chapter.

With regard to radiofrequency electromagnetic fields, the conclusions of an IARC Working Group of 15 countries and 30 members (including the author) were given at the carcinogenicity assessment conference on microwaves held by the IARC on May 24–31, 2011, and were essentially as follows.

1. Epidemiological: In its comprehensive summary of the study results that had been obtained so far, the Working Group concluded that there was limited evidence in humans of microwave carcinogenicity in its judgment based on the above-described reports of positive results.
2. Animal experiments: In its comprehensive summary of the study results that had been obtained so far, the Working Group noted that although most study results were negative, the above-described positive results in studies for combined carcinogenicity provided some evidence of carcinogenicity, and concluded that there was limited evidence in experimental animals of microwave carcinogenicity.
3. Cellular experiments: The comprehensive conclusion of the Working Group was that, although some studies showed positive results, the mechanistic evidence of microwave carcinogenicity was weak.
4. Summary: The human epidemiological studies and the experimental animal carcinogenicity studies were both deemed to provide limited evidence. The overall determination by the Working Group for the carcinogenicity by exposure to radiofrequency electromagnetic fields, including what it deemed the weak mechanistic evidence provided by the cell studies, was Group 2B (possibly carcinogenic to humans).

This Group 2B classification of radiofrequency electromagnetic fields represents nothing other than a determination of limited evidence of an association between electromagnetic fields from cellphones and brain tumors. An overview of this result was reported as a preliminary report [18]. The final report took two years or more from the end/conclusion of the conference to complete. The complete content consisted of 102 monographs and was issued in 2013 [28]. Figure 10.5 shows the IARC monograph for ELF, the WHO-EHC for ELF, and the IARC monograph for RF. About the same time as their issuance, the chair of this assessment committee and individuals associated with the IARC presented independent comments on cellphone carcinogenicity assessment [29], in which they emphasized the essential need for continuing provision of information to the public.

The number of categories assessed for carcinogenicity by the IARC has now reached 1001, and include chemicals in daily life and workplace environments, ionizing radiation, ultra violet, electromagnetic fields, foods and beverages and their ingredients, medicines, work environments, and air

IARC Monograph Volume 80   WHO EHC N° 238   IARC Monograph Volume 102
(2002)   (2007)   (2013)

Figure 10.5 Monograph and Environmental Health Criteria (EHC) for a static and an extremely low frequency electromagnetic field (ELF) carcinogenicity assessment from IARC (*left*), ELF-EHC health assessment from WHO (*center*), and a radiofrequency electromagnetic field (RF) carcinogenicity assessment from IARC (*right*).

environment [30]. Table 10.3 shows the total number and representative examples of the categories assessed and classified as of 13 April, 2017.

Following its receipt of the IARC radiofrequency electromagnetic field carcinogenicity assessment, the WHO is scheduled to proceed with comprehensive assessment of carcinogenicity and other health effects in conjunction with formulation of its Environmental Health Criteria (EHC) in 2017 or 2018. It issued a draft EHC [31] on September 30, 2014 and invited public comments on the draft until December 15 of that year. Chapters 2 to 12 of the fourteen EHC chapters were made public in the draft. Chapters 1 (overview and recommended research themes), 13 (health risk assessments), and 14 (protective measures) are important but their preparation must await completion of the Task Group conclusions, which will include consideration of the public comments.

On the other hand, the Scientific Committee on Emerging and Newly Identified Health Risks (SCENIHR), an affiliate of the European Commission, presented its *Opinion on Potential health effects of exposure to electromagnetic fields* on January 20, 2015 [32]. Its conclusions on the health effects of radiofrequency electromagnetic fields may be briefly summarized as follows.

Table 10.3 Typical IARC carcinogenicity classifications

Classification and Classification Criteria of Carcinogenicity	Results of Classification [1001 Cases]
Group 1: Carcinogenic to humans	Asbestos, Cadmium and cadmium compounds, Formaldehyde, *γ-rays, *Solar radiation, *X-rays, Alcoholic beverages, Coal-tars, Involuntary smoking, Tobacco smoking, *Ultraviolet radiation (wavelengths 100–400 nm, encompassing UVA, UVB, and UVC), *Sunlamps and sunbeds (see Ultraviolet-emitting tanning devices) [120 cases including others]
Group 2A: Probably carcinogenic to humans	Acrylamide, Adriamycin, Benz[a]anthracene, Benzo[a]pyrene, Cisplatin, Methyl methanesulfonate, Diesel engine exhaust, Polychlorinated biphenyls, [81 cases including others]
Group 2B: Possibly Carcinogenic to humans	Acetaldehyde, AF-2, Bleomycins, Chloroform, Daunomycin, Lead, *Magnetic fields (extremely *low-frequency), Merphalan, Methylmercury compounds, Mitomycin C, Phenobarbital, *Radiofrequency electromagnetic fields, Coffee (urinary bladder), Gasoline, [294 cases including others]
Group 3: Unclassifiable as to carcinogenicity to humans	Actinomycin D, Ampicillin, Anthracene, Benzo[e]pyrene, Cholesterol, Diazepam, *Electric fields (extremely low-frequency), *Electric fields (static), Ethylene, *Fluorescent lighting, *Magnetic fields (static), 6-Mercaptopurine, Mercury, Methyl chloride, Phenol, Toluene, Xylenes, Tea [505 cases including others]
Group 4: Probably not carcinogenic to humans	Caprolactam (material of nylon) [1 case]

*Environmental factors in relation to elelctromagnetic fields, ultraviolet, and ionizing radiation.

- The epidemiological study results do not provide sufficient evidence to indicate an increased risk of brain tumors, nor do they indicate an increased risk of other craniocervical cancers (including those of children and other malignancies).
- The results of some early studies posed questions concerning an increased risk of glioma and acoustic neuroma in heavy users of cell-phones. The indication of increased glioma risk has been weakened by the results of proximate cohort studies and studies of time-dependent incidence rates, but the possibility of an association between acoustic neuroma and radiofrequency exposure remains unresolved.

- The results of studies which suggested that radiofrequency exposure may affect waking and sleeping electroencephalograms have been further substantiated by more recent studies, but the biological significance of the small physiological changes is unclear.
- There is no evidence that radiofrequency exposure affects human cognitive functions.
- The earlier SCENIHR Opinion conclusion that radiofrequency electromagnetic field exposure levels below the current exposure limits have no harmful effect on reproduction or development remains unchanged by inclusion of recent data.

10.5 Electromagnetic Hypersensitivity

During the past ten years or more, the number of people around the world who attribute a decline in their physical or mental condition to a heightened sensitivity to electromagnetic fields has been increasing. They are referred to as having "electromagnetic hypersensitivity", in the mass media and elsewhere and more precisely as having "electrical (electromagnetic) hypersensitivity; EHS" by the WHO. It is said that skin manifestations (e.g., reddening, burning sensation) or autonomic disorders (e.g., headache, fatigue, dizziness, nausea) occur under exposure to weak Electromagnetic fields. Occurrence is not attributed to any specific frequency band, and seems rather to be attributable to both high and low frequencies.

Some hospitals in Europe and the U.S. began to provide care for hypersensitivity patients of this category in the late 1990s, and the number of such patients seems particularly high in northern Europe. The WHO convened an EHS workshop Prague, Czech Republic in 2004, and has also issued an EHS factsheet [33]. EHS is held to differ in various ways from chemical hypersensitivities (e.g., sick house syndrome), but no association of electromagnetic fields has been found in blind tests (patients uninformed of test-related electromagnetic fields exposure) of patients with subject symptoms for a causal relation. The WHO's current position based on scientific data is negative for EHS as an effect of electromagnetic fields.

10.6 Biological Effects of Electromagnetic Fields and Risk Communication

As noted, modern society moves energy far and wide in the form of electricity. The role of multifarious electromagnetic fields in the environment for information communication and other purposes has become extremely large and its growth will no doubt accelerate in the years ahead. This brings added convenience, but it has also given rise to apprehension among many people about electromagnetic fields and in particular unease about their health effects. In this light, many members from many countries in the WHO Task Group pointed out the importance of risk communication. The electromagnetic fields considered by the Task Group were non-ionizing low- and radiofrequency electromagnetic fields, unlike the ionizing X-rays and gamma rays that are commonly associated with the term "radiation" by the public. Although the energy levels of electromagnetic fields make it highly unlikely that they can directly damage cellular DNA, for large segments of the public the term "electromagnetic field" may hold the same connotations as "radiation". Related government organs in Japan such as the Ministry of Economy, Trade and Industry, the Ministry of Internal Affairs and Communications, and the Ministry of the Environment are using their websites and other means to increase public knowledge in this regard, and a policy is in progress of holding lectures on electromagnetic fields and health, describing the current state of knowledge to more people, and deepening understanding. At the same time, there are many publications and websites that promote excessive unease and anxiety over electromagnetic fields.

Risk communication is essential for fostering a clear understanding of electromagnetic fields and health. Without further research and progress in the life sciences sector concerning currently uncertain aspects, however, its effectiveness will be limited. Concurrent progress may be regarded as essential.

10.7 Conclusion

A primary objective of bioelectromagnetics is valid assessment of the biological effects of Electromagnetic fields based on the findings of scientifically reliable research. Research findings on the response of organisms, cells, and polymeric compounds to magnetic flux densities far higher than those in the usual environment are important, as they can lead to future advances in this field. They will enable derivation of threshold values based on an electromagnetic field dose-effect relation (at present, dose is based on magnetic flux density and induced current for low-frequency fields and specific absorption rate for radiofrequency electromagnetic fields, with exposure time as an added factor). Research is also in progress for the use of electromagnetic fields as a research tool in the life sciences themselves and for active application of the known effects of electromagnetic fields in the life sciences to fields of engineering, agriculture, health, and medicine.

Advances in cellphones and wireless power supply drive and in turn are being driven by rapid progress in engineering technology. At the same time, electromagnetic fields are a new environmental factor that must be considered as a part of society. Electromagnetic fields are non-ionizing, but it is highly likely that much of the public will tend to identify the term "electromagnetic fields" with "ionizing radiation" as it is commonly understood. In the near future, the use of electromagnetic fields in conjunction with the emergence of cellphone and computer wireless batteries, electric vehicles wireless power supply, and other noncontact energy transmission technologies will only follow a rapidly rising trajectory. The inexorable rise of the electromagnetic environment will require further research for determination of electromagnetic field safety based on firm scientific data for the unexplained aspects and for advancement of leading-edge technologies in the life sciences.

References

[1] Kato, M., Shigemitsu, T., Miyakoshi, J., Fujiwara, O., Wang, J., Yamazaki, K., et al. (2006). *Electromagnetics in Biology*. Berlin: Springer.

[2] Miyakoshi, M., Schoemaker, M. J., Preece, A. W., Leitgeb, N., Bernardi, P., Lin, J. C., Lin, J. C. (2009). "Health effects of cell phone radiation," in *Advances in Electromagnetic Fields in Living Systems*, Vol. 5, ed. J. C. Lin (New York: Springer).

[3] Miyakoshi, M. (2013). Cellular and molecular responses to radio-frequency electromagnetic fields. *Proc. IEEE* 101, 1494–1502.

[4] Wertheimer, N., Leeper, E. D. (1979). Electrical wiring configurations and childhood cancer. *Am. J. Epidemiol.* 109, 273–284.

[5] Kheifets, L., Shimkhada, R. (2005). Review; childhood leukemia and EMF: review of the epidemiologic evidence. *Bioelectromagnetics* 7, S51–S59.

[6] Kabuto, M., Nitta, H., Yamamoto, S., Yamaguchi, N., Akiba, S., Honda, Y., et al. (2006). Childhood leukemia and magnetic fields in Japan: a case-control study of childhood leukemia and residential power-frequency magnetic fields in Japan. *Int. J. Cancer* 119, 643–650.

[7] Ahlbom, A., Day, N., Feychting, M., Roman, E., Skinner, J., Dockerty, J., et al. (2000). A pooled analysis of magnetic fields and childhood leukaemia. *Br. J. Cancer* 83, 692–698.

[8] Kheifets, L., Ahlbom, A., Crespi, C. M., Draper, G., Hagihara, J., Lowenthal, R. M., et al. (2010). Pooled analysis of recent studies on magnetic fields and childhood leukaemia. *Br. J. Cancer* 103, 1128–1135.

[9] World Health Organization (2017). Available at: http://www.iarc.fr/en/media-centre/pr/2010/pdfs/pr200_E.pdf#search ='IARCWHO Press Release No. 200'

[10] Hardell, L., Carlberg, M., and Hansson Mild, K. (2011). Pooled analysis of case-control studies on malignant brain tumours and the use of mobile and cordless phones including living and deceased subjects. *Int. J. Oncol.* 38, 1465–74.

[11] Sato, Y., Akiba, S., Kubo, O., and Yamaguchi, N. (2011). A case-case study of mobile phone use and acoustic neuroma risk in Japan. *Bioelectromagnetics* 32, 85–93.

[12] Aydin, D., Feychting, M., Schz, J., Tynes, T., Andersen, T. V., Schmidt, L. S., et al. (2011). Mobile phone use and brain tumors in children and adolescents: a multicenter case–control study. *Natl. Cancer Inst.* 103, 1–13.

[13] Cardis, E., Armstrong, B. K., Bowman, J. D., Giles, G. G., Hours, M., Krewski, D., et al. (2011). Risk of brain tumours in relation to estimated RF dose from mobile phones—results from five Interphone countries. *Occup. Environ. Med.* 68, 631. DOI:10.1136/oemed-2011-100155

[14] Mobi kids (2015). Available at: http://www.crealradiation.com/index.php/en/mobi-kids-home

[15] GERoNiMO (2016). Available at: http://www.crealradiation.com/index.php/en/geronimo-home

[16] IARC (2002). *Monograph on the Evaluation of Carcinogenic Risks to Humans,* Vol. 80. IARC: Lyon.

[17] Repacholi, M. H., Basten, A., Gebski, V., Noonan, D., Finnie, J., and Harris, A. W. (1997). Lymphomas in E-Piml transgenic mice exposed to pulsed 900 MHz electromagnetic fields. *Radiat. Res.* 147, 631–640.

[18] Baan, R., Grosse, Y., Lauby-Secretan, B., El Ghissassi, F., Bouvard, V., Benbrahim-Tallaa, L., et al. (2011). Carcinogenicity of Radiofrequency electromagnetic fields. *Lancet Oncol.* 12, 624–626.

[19] Szmigielski, S., Szudzinski, A., Pietraszek, A., Bielec, M., Janiak, M., Wrembe, J. K. (1982). Accelerated development of spontaneous and benzopyrene-induced skin cancer in mice exposed to 2450-MHz microwave radiation. *Bioelectromagentics* 3, 179–191.

[20] Tillmann, T., Ernst, H., Streckert, J., Zhou, Y., Taugner, F., Hansen, V., et al. (2010). Indication of cocarcinogenic potential of chronic

UMTS-modulated radiofrequency exposure in an ethylnitrosourea mouse model. *Int. J. Radiat. Biol.* 86, 529–541.

[21] Heikkinen, P., Ernst, H., Huuskonen, H., Komulainen, H., Kumlin, T., Mki-Paakkanen, J., et al. (2006). No effects of radiofrequency radiation on 3-chloro-4-(dichloromethyl)-5-hydroxy-2(5H)-franone-induced tumorigenesis in female Wister rats. *Radiat. Res.* 166, 397–408.

[22] Wyde, M. (2016). *NTP Toxicology and Carcinogenicity Studies of Cell Phone Radiofrequency Radiation.* Ghent: BioEM.

[23] Crasta, K., Ganem, N. J., Dagher, R., Lantermann, A. B., Ivanova, E. V., Pan, Y., et al. (2012). DNA breaks and chromosome pulverization from errors in mitosis., *Nature* 482, 53–58. doi: 10.1038/nature10802

[24] Leszczynski, D., Joenvr, S., Reivinen, J., Kuokka, R. (2002). Non-thermal activation of the hsp27/p38MAPK stress pathway by mobile phone radiation in human endothelial cells: molecular mechanism for cancer- and blood-brain barrier-related effects. *Differentiation* 70, 120–129.

[25] World Health Organization (2014). Available at: http://www.who.int/ peh-emf/project/en/

[26] World Health Organization (2007). Available at: http://www.who.int/ peh-emf/publications/extremely-low frequencyelectromagnetic field_ ehc/en/index.html

[27] World Health Organization (2008). *Extremely Low Frequency Fields-Environmental Health Criteria No.238.* Geneva: WHO.

[28] IARC Working Group on the Evaluation of Carcinogenic Risks to Humans (2013). Non-Ionizing Radiation, Part 2: Radiofrequency Electromagnetic Fields. *IARC Monogr. Eval. Carcinog. Risks Hum.* 102, 1–460.

[29] Samet, J. M., Kurt, S., Joachim, S., Rodolfo, S. ()2014. Mobile phones and cancer –next steps after the 2011 IARC review. *Epidemiology* 25, 23–27.

[30] World Health Organization (2011). Available at: http://monographs.iarc. fr/ENG/Classification/

[31] World Health Organization (2014). Available at: http://www.who.int/ peh-emf/research/radiofrequency _ehc_page/en/

[32] European Commission (2017). Available at: http://ec.europa.eu/health/ scientintermediate-frequencyic_committees/emerging/opinions/index_ en.htm

[33] World Health Organization (2014). Available at: http://www.who.int/ mediacentre/factsheets/fs296/en/index.html

11

Coexistence of WPT and Wireless LAN in a 2.4-GHz Band[*]

Koji Yamamoto and Shota Yamashita

Graduate School of Informatics, Kyoto University, Japan

Abstract

The 2.4-GHz frequency band is a good candidate for wireless power transmission (WPT) [1]. Because numerous wireless local area network (WLAN) devices use the 2.4-GHz band, coexistence between WPT and WLAN should be carefully considered. In this chapter, the coexistence problem is experimentally discussed, particularly WLAN devices powered by WPT. In general, two schemes exist to divide radio resources: frequency and time divisions. Radio resource management to enable coexistence of WPT and WLAN in the frequency domain is referred to as *adjacent channel operation* of continuous WPT and WLAN data transmission. That in the time domain is referred to as *co-channel operation* of intermittent WPT and WLAN data transmission. Experimental results reveal that, even when these operations are implemented, several new problems arise because WLAN devices have been developed without considering the existence of WPT. One problem clarified in our experiment is that adjacent channel operation in a 2.4-GHz band does not necessarily perform well because of the considerable interference. This interference occurs regardless of the frequency separation in a 2.4-GHz band. The other problem is that intermittent WPT may result in throughput degradation because of rate adaptation in WLAN. As one solution

---

[*]This chapter is based on [2] copyright©2014 IEICE, permission number: 17RA0003, [3] copyright©2014 IEICE, permission number: 17RA0004, and [4] copyright©2015 IEICE, permission number: 17RA0014.

to this throughput degradation, an exposure assessment-based rate adaptation scheme is implemented and evaluated.

11.1 Introduction

WPT for wireless devices has been discussed in several previous studies; [5, 6] designed antennas and rectifiers, and [7, 8] showed that mobile phones and sensor devices could be powered wirelessly using the 2.4-GHz band. In [9], the authors investigated the effects of WPT on IEEE 802.15-based communications. However, in these previous studies, the effects of WPT on WLAN communications were not discussed in detail.

The main purpose of this chapter is to investigate experimentally the impact of WPT on IEEE 802.11g-based WLAN data communications. In particular, we assume that a wireless device is powered wirelessly and that the device must transmit its data using stored energy.

In this study, WPT is assumed to be conducted at the same 2.4-GHz with WLAN for efficient use of the spectrum. Note that the 2.4-GHz frequency band is one candidate for wireless power transmission [1].

Two fundamental methods exist to avoid interference from WPT to WLAN data communications without modifying WLAN devices: 1) *adjacent channel operation* of continuous power and data transmissions (i.e., the frequency-domain separation) and 2) *co-channel operation* of intermittent microwave power and data transmissions (i.e., the time-domain separation). These methods are implemented and their performance is evaluated.

The results of our experiment reveal that several problems occur that cannot be solved even using these fundamental methods. This is because WLAN devices have been developed without considering WPT. Adjacent channel WLAN devices would be affected by WPT because they have been developed without considering the amount of power from a microwave energy source. However, when WPT is conducted intermittently, WLAN devices are expected to communicate successfully because of the carrier sense multiple access with collision avoidance (CSMA/CA) mechanism.

This chapter is organized as follows. Sections 11.2 and 11.3 describe the adjacent channel operation and co-channel operations, respectively. Section 11.4 discusses the exposure assessment-based rate adaptation scheme to avoid throughput degradation in the co-channel operation. Section 11.5 summarizes the main findings.

11.2 Adjacent Channel Operation of Continuous WPT and WLAN Data Transmission

In this section, to discuss the feasibility of adjacent channel operation of continuous WPT and WLAN data transmission, we experimentally measure the throughput of WLAN. Specifically, we evaluate both the impact of received power density at a WLAN device and that of the frequency separation on the throughput. Experimental results reveal that even when using adjacent channel operation, WLAN devices are affected by WPT. This is because the received WPT power is considerable and thus cannot be attenuated by using a band-pass filter.

11.2.1 Experimental Setup for Continuous WPT

Figure 11.1 shows the setup for measurements. The proposed system consists of an energy source (ES), data transmitter (DT), and data receiver (DR). Figure 11.2 shows the ES and the DT used in this experiment. The DR was positioned behind the ES to avoid any effect from the WPT. All measurements presented in this chapter were performed in a radio-anechoic chamber.

The ES was composed of a signal generator, an amplifier, and a horn antenna. The ES transmits continuous microwaves to the DT. Note that the use of continuous microwaves for WPT is a general assumption [6, 7, 10, 11]. Through measurements, we confirmed that the bandwidth for WPT was less than 2 kHz. The center frequency, f_{WPT}, was set to fall within a 2.4–2.5 GHz band.

As the DT, a laptop PC (Apple MacBook Pro) was used. The DT transmitted user datagram protocol (UDP) data frames to the DR using Iperf 2.0.5 [12]

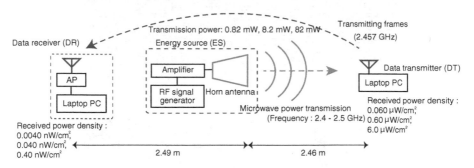

Figure 11.1 Setup of adjacent channel operation.

for 20 s at a center frequency of 2.457 GHz. The offered load and the UDP datagram size were 15 Mbit/s and 1470 B, respectively.

The DR consisted of an access point (Allied Telesis AT-TQ2403) and a laptop PC. Using Iperf, the laptop PC on the DR received data frames from the DT via the access point (AP) and then measured the throughput.

11.2.2 Measurement Results

Figure 11.3 shows the achieved throughput. The throughput characteristics depend highly on the received power density at the DT. When the received power density at the DT was $\leq 0.60\,\mu\text{W/cm}^2$, a throughput of approximately 15 Mbit/s was achieved unless the center frequency of the WPT, f_{WPT}, overlapped the channel for the WLAN (i.e., the DT did not detect WPT from the ES). Here, recall that the offered load was 15 Mbit/s. By contrast, when f_{WPT} overlapped the channel for WLAN communications, the throughput was almost zero (i.e., the DT detected WPT and deferred its transmission).

However, when the received power density at the DT was $6.0\,\mu\text{W/cm}^2$, the throughput was zero regardless of f_{WPT}. This may be because the band-pass filter at the receiver did not completely attenuate the WPT energy.

If WLAN devices are powered wirelessly, the received power density at the devices generally must be much greater than $6.0\,\mu\text{W/cm}^2$. In this case, the WLAN modules would detect microwave energy from the ES regardless of the value of f_{WPT}. For example, as estimated in [3], the WLAN sensor node must receive a power density of $0.3\,\text{mW/cm}^2$, even in sleep mode. At

Figure 11.2   Experimental setup of the ES and DT.

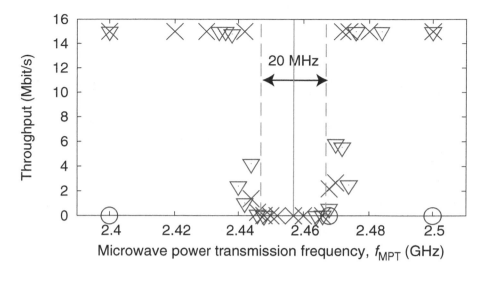

Figure 11.3 Throughput of adjacent channel operation.

least two possible solutions to this power-related problem exist. The first is to use an appropriate band-pass filter at the receiver to attenuate the microwave power of WPT. The second is to transmit microwave power intermittently. Because the second solution can be implemented by using commercially available devices, in this chapter, we discuss intermittent WPT.

11.3 Co-Channel Operation of Intermittent WPT and WLAN Data Transmissions

In this section, we discuss the feasibility of the co-channel operation of intermittent WPT and WLAN data transmission. Because WLAN devices are operated based on CSMA/CA, WLAN devices defer their transmission during WPT and thus they are expected to communicate successfully, even when WPT is conducted intermittently. To assess the feasibility of the co-channel operation in detail, we measure the data rate and the number of discarded data frames.

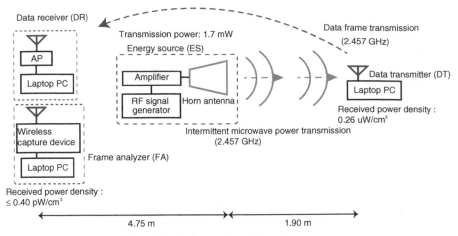

Figure 11.4   Setup of co-channel operation.

11.3.1 Experimental Setup for Intermittent WPT

Figure 11.4 shows the setup of the experiment, which consists of the ES, DT, DR, and a frame analyzer (FA). The DR and FA were positioned behind the ES to avoid any effect from WPT. The FA consisted of a WLAN frame capture device and a laptop PC, and it captured the frames from the DT and DR. The motivation of frame capturing is to investigate the data rate that is specified in the header of each frame as well as the number of transmitted frames.

The ES transmits microwave power periodically. Specifically, it conducts WPT during a fixed time, T_{PT}, and stops WPT during the other fixed time, T_{PS}. Note that the subscript "PT" represents *power transmission*, and the subscript "PS" represents *power suspension*. The ES is set to transmit at a center frequency of 2.457 GHz and transmission power of 1.70 mW.

During T_{PT}, the DT detects WPT energy from the ES. Thus, the DT does not transmit data frames and instead holds them in a finite-size buffer during T_{PT}. In T_{PS}, the DT transmits them. As mentioned in Section 11.2, the offered load and UDP datagram size were 15 Mbit/s and 1470 B, respectively.

11.3.2 Estimation of Frame Loss Rate of Co-channel Operation

According to CSMA/CA procedure, WLAN devices are expected to detect WPT and defer their transmissions until the end of WPT. In addition, if the output buffer of WLAN devices is filled, according to a pre-programmed buffer management rule [13], WLAN devices will delete particular frames in the buffer. In this section, to differentiate the cause of frame losses, we clarify the condition necessary to avoid frame losses that occur because of buffer overflow, assuming tail-drop.

The frame loss rate P_{loss} is defined as:

$$P_{\text{loss}} := \frac{N_{\text{generated}} - N_{\text{received}}}{N_{\text{generated}}}, \tag{11.1}$$

where N_{received} and $N_{\text{generated}}$ represent the number of received and generated frames, respectively.

The output buffer on the DT overflows mainly when either T_{PT} is long or T_{PS} is short. To confirm these hypotheses and estimate the conditions to prevent frame losses, we formulated the frame loss rate as a function of T_{PT} and T_{PS}. For the sake of simplicity, the DT was assumed not to transmit data frames during T_{PT}.

First, when T_{PS} is sufficiently long, whether the output buffer overflows depends on the value of T_{PT}. Intuitively, that the output buffer overflows when the sum of the size of the data frames generated during T_{PT} is greater than the output buffer size B seems obvious. Let the offered load be denoted by G. The condition to prevent buffer overflow is then:

$$GT_{\text{PT}} \leq B. \tag{11.2}$$

Here, the equality of (11.2) holds if

$$T_{\text{PT}} = B/G =: T_{\text{PT,longPS}}. \tag{11.3}$$

Therefore, when T_{PS} is sufficiently long and $T_{\text{PT}} > T_{\text{PT,longPS}}$, the number of discarded frames, $N_{\text{generated}} - N_{\text{received}} =: N_{\text{discarded}}$, is calculated as:

$$N_{\text{discarded}} = \frac{GT_{\text{PT}} - B}{L}, \tag{11.4}$$

where L represents the UDP payload size.

Second, when $T_{\text{PT}} \leq T_{\text{PT,longPS}}$, whether the output buffer overflows depends on the ratio of T_{PS} to T_{PT}. When the number of data frames generated in a cycle is greater than the number of data frames that can be transmitted in T_{PS}, the buffer would overflow. Thus, the condition to prevent buffer overflow is

$$\frac{G(T_{\text{PT}} + T_{\text{PS}})}{L} \leq \frac{T_{\text{PS}}}{\tau}, \tag{11.5}$$

where τ represents the average value of the periods from the start of one data transmission to the next. Note that the value of τ is evaluated experimentally. Here, the equality of (11.5) holds if

$$T_{\text{PS}} = \frac{GT_{\text{PT}}}{L/\tau - G} =: T_{\text{PS,shortPT}}. \tag{11.6}$$

Therefore, when $T_{\text{PT}} \leq T_{\text{PT,longPS}}$ and $T_{\text{PS}} < T_{\text{PS,shortPT}}$, the value of $N_{\text{discarded}}$ is calculated as follows:

$$N_{\text{discarded}} = \frac{G(T_{\text{PT}} + T_{\text{PS}}) - LT_{\text{PS}}/\tau}{L}. \tag{11.7}$$

Third, when $T_{\text{PT}} > T_{\text{PT,longPS}}$, whether $N_{\text{discarded}}$ is calculated using (11.4) or (11.7) depends on the value of T_{PS}. In this case, the DT should transmit $(B + GT_{\text{PS}})/L$ of data frames during T_{PS}. Therefore, when

$$\frac{B + GT_{\text{PS}}}{L} \leq \frac{T_{\text{PS}}}{\tau}, \tag{11.8}$$

$N_{\text{discarded}}$ is calculated using (11.7). Here, the equality of (11.8) holds if

$$T_{\text{PS}} = \frac{B}{L/\tau - G} =: T_{\text{PS,longPT}}. \tag{11.9}$$

Consequently, by using equations from (11.2)-(11.6) and $N_{\text{generated}} = G(T_{\text{PT}} + T_{\text{PS}})/L$, the frame loss rate P_{loss} can be calculated as:

$$P_{\text{loss}} = \begin{cases} \dfrac{G(T_{\text{PT}} + T_{\text{PS}}) - LT_{\text{PS}}/\tau}{G(T_{\text{PT}} + T_{\text{PS}})}, & (11.10a) \\ \quad T_{\text{PS}} < T_{\text{PS,shortPT}} \text{ or } T_{\text{PS}} < T_{\text{PS,longPT}}; \\ \dfrac{GT_{\text{PT}} - B}{G(T_{\text{PT}} + T_{\text{PS}})}, & (11.10b) \\ \quad T_{\text{PT}} > T_{\text{PT,longPS}} \text{ and } T_{\text{PS}} \geq T_{\text{PS,shortPT}}; \\ 0, & (11.10c) \\ \quad T_{\text{PT}} \leq T_{\text{PT,longPS}} \text{ and } T_{\text{PS}} \geq T_{\text{PS,shortPT}}. \end{cases}$$

Note that P_{loss} is an increasing function of T_{PT} and a decreasing function of T_{PS}.

To avoid buffer overflow, $T_{\text{PT}} \leq T_{\text{PT,longPS}}$ and $T_{\text{PS}} \geq T_{\text{PS,shortPT}}$ are required. Thus, the ratio of T_{PS} to T_{PT} should satisfy

$$T_{\text{PS}}/T_{\text{PT}} \geq T_{\text{PS,shortPT}}/T_{\text{PT,longPS}}$$
$$= \frac{1}{L/G\tau - 1}. \tag{11.11}$$

In other words, the lower bound of $T_{\text{PS}}/T_{\text{PT}}$ must be increased along with the offered load G. Note that a larger value of $T_{\text{PS}}/T_{\text{PT}}$ means more time for data transmission and less for WPT.

The existence of the lower bound for $T_{\text{PS}}/T_{\text{PT}}$ means that an upper bound exists for the supplied power. Let the supplied power during T_{PT} be denoted by P_{PT}. The average supplied power over multiple cycles, P_{e}, can then be written as:

$$P_{\text{e}} = \frac{P_{\text{PT}}T_{\text{PT}}}{T_{\text{PT}} + T_{\text{PS}}} = \frac{P_{\text{PT}}}{1 + T_{\text{PS}}/T_{\text{PT}}}$$
$$\leq \frac{P_{\text{PT}}}{1 + T_{\text{PS,shortPT}}/T_{\text{PT,longPS}}}$$
$$= (1 - G\tau/L)P_{\text{PT}} =: P_{\text{e,max}}. \tag{11.12}$$

Thus, the greater offered traffic G means lower average supplied power. Figure 11.5 shows the relation between $P_{\text{e,max}}/P_{\text{PT}}$ and G. Here, $P_{\text{e,max}}/P_{\text{PT}}$ decreases as G increases. Therefore, G must be decreased to supply sufficient power to a WLAN device.

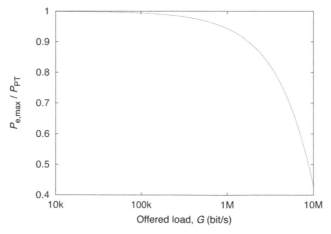

Figure 11.5 $P_{\text{e,max}}/P_{\text{PT}}$ versus offered load G for $\tau = 0.67\,\text{ms}$ and $L = 1470\,\text{B}$.

11.3.3 Measurement Results

11.3.3.1 Data rate

Figure 11.6 shows the average data rate specified in each frame for every 0.050 s. Note that this does not represent the achieved throughput. We can see that the data rate decreased during T_{PT} according to rate adaptation.

Specifically, in Figure 11.6, two factors exist that we did not considered in Section 11.3.2. These factors are related to rate adaptation; they increase the frame loss rate more than that estimated in (11.10). The first factor is that the DT attempts to retransmit data frames at a decreased data rate even during T_{PT}, as shown in Figure 11.6. This is because the DT sometimes attempts to transmit even during T_{PT} and fails to receive an acknowledgment (ACK) frame. In particular, the data rate gradually decreases along with T_{PT}.

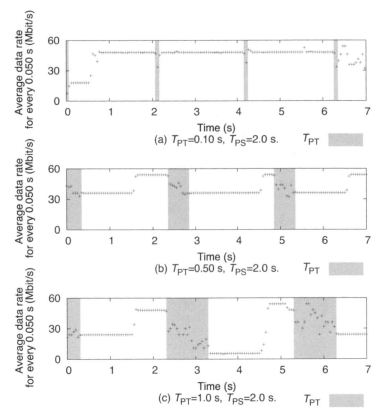

Figure 11.6   Average data rate for every 0.050 s in which: (a) $T_{PT} = 0.10$ s and $T_{PS} = 2.0$ s, (b) $T_{PT} = 0.50$ s and $T_{PS} = 2.0$ s, and (c) $T_{PT} = 1.0$ s and $T_{PS} = 2.0$ s.

The second factor is that a certain amount of time is required for the DT to reconfigure the data rate after WPT is stopped. In general, as shown in Figure 11.6(a), the DT generally transmits data frames at 48 Mbit/s during T_{PS}. However, as can be seen in Figs. 11.6(b) and 11.6(c), when T_{PT} is sufficiently long, the DT does not reconfigure the data rate even after the ES stops WPT.

One solution is to control the DT such that it does not transmit data frames during WPT. To achieve this, the ES and DT must share information on the timings of WPT and data transmission to enable this kind of control. Another solution is to differentiate the case of frame losses. This approach is discussed in Section 11.4.

11.3.3.2 Frame loss rate

Figure 11.7 shows the frame loss rate versus T_{PS} for $T_{PT} = 0.50$ s and $T_{PT} = 1.0$ s. The frame loss rate P_{loss} decreases as T_{PS} increases, as can be easily understood from (11.10). We can see that (11.10a) and (11.10b) are appropriate models to establish the relation between P_{loss} and T_{PS} when $T_{PT} = 1.0$ s in the range $T_{PS} \geq 2.0$ s. In this range, $\tau = 0.65$ ms is measured

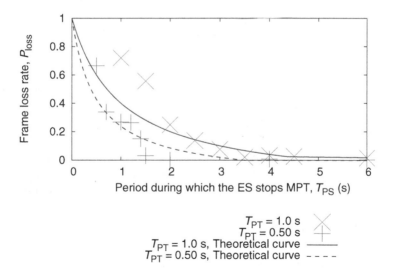

Figure 11.7 Frame loss rate versus period during which the ES suspends WPT. The solid line represents the fitting of the experimental data of $T_{PT} = 1.0$ s to estimations of the frame loss rate, provided $B = 1.6$ MB and $\tau = 0.65$ ms. The dashed line represents the fitting of the experimental data of $T_{PT} = 0.5$ s to estimations of the frame loss rate, provided $B = 1.6$ MB and $\tau = 0.68$ ms.

independent of T_{PS}, and B is estimated by fitting the experimental data $T_{PT} = 1.0\,s$ in this range to the estimations (11.10) using a least-squares method. Therefore, the estimations of the frame loss rate are calculated as follows:

$$P_{\text{loss}} = \begin{cases} \dfrac{1.2\,s}{T_{PS} + 1.0\,s} - 0.21, & T_{PS} < 4.4\,s; \\ \dfrac{0.15\,s}{T_{PS} + 1.0\,s}, & T_{PS} \geq 4.4\,s. \end{cases} \tag{11.13}$$

Note that $T_{PS,\text{longPT}}$ is calculated as $4.4\,s$ using (11.9). In the other range (i.e., when $T_{PT} = 1.0\,s$ and $T_{PS} \leq 1.5\,s$), the estimations do not fit the experimental data well. This is a result of the first and second factors that we have not considered in Section 11.3.2. In particular, the second factor is the main cause. Recall that, when T_{PS} is short, the period during which the DT transmits at a low rate increases.

We can see that (11.10a) and (11.10c) are appropriate models to establish the relation between P_{loss} and T_{PS} when $T_{PT} = 0.50\,s$ in the range $T_{PS} \geq 0.70\,s$. In this range, $\tau = 0.69\,ms$ is measured independent of T_{PS}. Therefore, the estimation of the frame loss rate is calculated as follows:

$$P_{\text{loss}} = \begin{cases} \dfrac{0.58\,s}{T_{PS} + 0.50\,s} - 0.15, & T_{PS} < 3.5\,s; \\ 0, & T_{PS} \geq 3.5\,s. \end{cases} \tag{11.14}$$

Note that $T_{PS,\text{shortPT}}$ is calculated as $3.5\,s$ using (11.6). In the other range (i.e., when $T_{PT} = 0.50\,s$ and $T_{PS} = 0.50\,s$), the estimations do not fit the experimental data well. This is also caused by the second factor that we have not considered in Section 11.3.2.

Note that the difference between the experimental data for $T_{PT} = 1.0\,s$ and the theoretical curve (11.13) is greater than that between the experimental data for $T_{PT} = 0.50\,s$ and the theoretical curve (11.14). This is mainly caused by the first factor that we have not considered in Section 11.3.2. Recall that the longer T_{PT} is, the more the DT fails to receive ACK frames. Thus, the data rate decreases.

Figure 11.8 shows the frame loss rate P_{loss} versus T_{PT} for $T_{PS} = 2.0\,s$ and $T_{PS} = 6.0\,s$. Here, T_{PT} is set to $\geq 0.10\,s$ in this experiment. This is because similar trends to that in which $T_{PT} = 0.10\,s$ would be achieved when $T_{PT} < 0.10\,s$. The frame loss rate P_{loss} increases along with T_{PT}, as can be easily understood from (11.10). When $T_{PS} = 2.0\,s$, the estimations (11.10) do not fit the experimental data. This is caused by the second factor

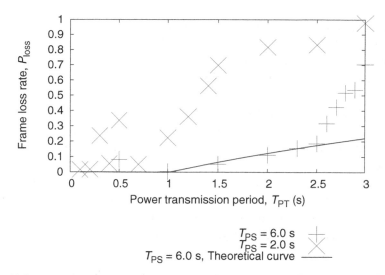

Figure 11.8 Frame loss rate versus power transmission period. The solid line represents the fitting of the experimental data of $T_{PS} = 6.0$ s to estimations of the frame loss rate, provided $B = 1.9$ MB and $\tau = 0.67$ ms.

previously mentioned. Recall that T_{PS} is so short that the data rate cannot be reconfigured during T_{PS}.

In addition, we can see that when $T_{PS} = 6.0$ s in the range $0 \leq T_{PT} \leq 2.5$ s, (11.10b) and (11.10c) are appropriate models to establish the relation between P_{loss} and T_{PT}. In this range, $\tau = 0.67$ ms is measured independent of T_{PT}, and B is estimated by fitting the experimental data in the range from 1.0 s $\leq T_{PT} \leq 2.5$ s to the estimation (11.10b) using a least-squares method. Therefore, when $T_{PS} = 6.0$ s, the estimations of frame loss rate (11.10b) and (11.10c) are calculated as follows:

$$
P_{loss} = \begin{cases} 1.0 - \dfrac{5.0\,\mathrm{s}}{T_{PT} + 6.0\,\mathrm{s}}, & T_{PT} > 1.0\,\mathrm{s}; \\ 0, & T_{PT} \leq 1.0\,\mathrm{s}. \end{cases} \tag{11.15}
$$

Note that $T_{PT,longPS}$ is calculated as 1.0 s using (11.3).

In the other range (i.e., $T_{PT} \geq 2.6$ s and $T_{PS} = 6.0$ s), the estimations (11.10a), (11.10b), and (11.10c) do not fit the experimental data well. This is caused by a factor that is not discussed in this study. Hereafter, we call it the third factor, in which the DT often fails to receive frames from the DR during T_{PT}, including beacon frames. We find that the DT attempts to go to sleep and does not transmit data for a certain time when $T_{\mathrm{PT}} \geq 2.6\,\mathrm{s}$. Thus, P_{loss} is higher than the estimation. Note that details related to sleep control are not standardized and depend on the device characteristics. In addition to these sleep periods, a network disassociation would be caused by beacon reception failure.

As an example of how to avoid these sleep periods as well as network disassociation, either microwave power should not be transmitted during a beacon transmission or some timer values related to beacon reception should be tuned.

These results reveal that the estimations in Section 11.3.2 match the experimental data well, excluding the range in which the frame loss rate is affected by any one of the three factors. Because of these factors, more data frames are discarded than expected in Section 11.3.2.

11.4 Exposure Assessment-Based Rate Adaptation

This section introduces rate adaptation based on exposure assessment [4]. The purpose of the rate adaptation scheme is to solve the data rate reduction problem described in Section 11.3.3.1, that is: exposure of the station to microwave radiation causes a lower physical (PHY) layer data rate to be selected, the use of which continues even after the microwave radiation is discontinued.

The main idea of the proposed scheme is to utilize the output of the rectenna. Using the rectenna output, a WLAN station can assess whether the station is exposed to microwave radiation. Then, using historical data corresponding to the assessment results, the station selects an appropriate PHY data rate. The historical data are obtained from previous transmission results (e.g., historical data pertaining to the data frame loss ratio).

The scheme was implemented and verified through an experiment. Experimental results showed that the proposed scheme prevents the reduction in the PHY data rate, which is caused by using historical data stored in a single memory. Thus, the proposed scheme improves WLAN throughput.

11.4.1 Rate Adaptation Schemes

In most rate adaptation schemes, by using data related to previous transmission results (e.g., historical data of the data frame loss ratio), the PHY data rate for data transmission is adjusted. The purpose of using historical data is to estimate current link quality, which depends on the distance between a data transmitter and data receiver as well as on interference power at the data receiver. To maintain link quality, the PHY data rate is adjusted so that data frames are successfully received by the data receiver. However, most conventional rate adaptation schemes are designed without assuming that high interference power at the station causes PHY data rate reduction.

Some previous studies have proposed rate adaptation schemes that are capable of assessing whether a WLAN station is exposed to microwave radiation from devices using Bluetooth, ZigBee, or from microwave ovens. SGRA [14] and ARES [15] attempt to assess whether a WLAN station is exposed to microwave radiation based on both the signal-to-noise power ratio (SNR) and the data frame loss ratio. However, because of the exposure assessment based on the data frame loss ratio, when the data receiver experiences strong SNR degradation, the station erroneously detects that it is exposed to microwave radiation for WPT even when this is not the case. This is caused by an increase in the data frame loss ratio, not only because of exposure of the station to microwave radiation for WPT but also because of SNR degradation at the data receiver. The increase in the data frame loss ratio with SNR degradation is experimentally demonstrated in [16].

11.4.2 Rate Adaptation Scheme Based on Exposure Assessment Using Rectenna Output

The design of the proposed scheme focuses on the use of historical data in conventional rate adaptation. Recall that historical data are obtained from previous transmissions (e.g., historical data of the data frame loss ratio). This proposed scheme has the following two features: (i) the station assesses whether it is exposed to microwave radiation for WPT by using rectenna output power; (ii) the station selects an appropriate PHY data rate using historical data that correspond to the assessment results. The reasons for using rectenna output power are two-fold. First, a rectenna has been installed in a station powered with WPT and thus no need exists to install other devices to perform the exposure assessment. Second, the use of rectenna output power enables a station equipped with a rectenna to assess directly whether it is exposed to microwave radiation for WPT.

Immediately before rate adaptation, the station measures the rectenna output power p_{o}. Then, by using p_{o}, the station assesses whether it is exposed to microwave radiation for WPT. Let the power threshold be denoted by P_{th}.

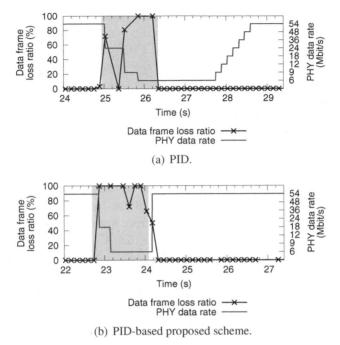

(a) PID.

(b) PID-based proposed scheme.

Figure 11.9   Data frame loss ratio and PHY data rate for each of the PID-based rate adaptation, where the gray area indicates the time during which the DT is exposed to microwave radiation for WPT.

By using the historical data in the memory corresponding to the assessment results, the station then selects an appropriate PHY data rate. The two independent memories in which the historical data for rate adaptation purposes are stored are denoted by M_E and M_{NE}, where the subscripts "E" and "NE" represent "exposure" and "non-exposure," respectively. When $p_o > P_{th}$, the station determines that it is exposed to microwave radiation for WPT, after which it selects an appropriate PHY data rate using the historical data stored in M_E. However, when $p_o \le P_{th}$, the station determines that it is not exposed to microwave radiation for WPT and then selects an appropriate PHY data rate using the historical data stored in M_{NE}.

Figure 11.9 shows both the data frame loss ratio and PHY data rate for each of proportional-integral-differential (PID)-based rate adaptation. The red lines in Figure 11.9 confirm that the PHY data rate is reduced when the DT is exposed to microwave radiation for WPT. When we compare the PHY data rate in Figure 11.9(a) with that in Figure 11.9(b), an improvement is clearly noticeable.

11.5 Conclusion

We experimentally clarified the requirements for both WPT and IEEE 802.11-based WLAN data transmissions using the same 2.4-GHz band. In particular, we identified three specific issues related to WPT that affect WLAN devices. In general, adjacent channel operation of microwave power and data transmissions may be a possible solution. However, we first demonstrated that nearly all data communications failed regardless of the frequency separation in the 2.4-GHz band when the supplied microwave power was sufficient for the WLAN devices. This may be because the band-pass filter at the receiver did not completely attenuate the microwave energy. Because numerous WLAN devices have already been used, solving this problem by only changing the frequency of WPT at 2.4 GHz is difficult.

Thus, we measured the frame loss rate of data transmissions during intermittent WPT. In general, if the offered load is set so that the output buffer of the WLAN devices does not overflow, frames would not be discarded because of CSMA/CA. However, we also determined that WLAN devices inefficiently decrease their data rate if they attempt to transmit during WPT. In addition, we showed that if a WLAN device does not receive consecutive beacons above a certain number because of WPT, it switches its mode to sleep, or it is disassociated from the AP.

Many possible solutions exist to resolve these issues. To avoid the data rate degradation, the following solutions seem to be effective: setting the DT such that it does not transmit data frames during WPT and does not reduce the data rate, and setting the DT to increase the data rate immediately after WPT is stopped. In addition, to avoid sleep periods and disassociation from the AP, the following solutions seem to be effective: tune the timing parameters related to sleep or disassociation and set the ES not to transmit microwave power while the DT receives beacon frames. In particular, to avoid interference from WPT to beacon receptions, the ES and DT are must share information on timings of WPT and data transmission. By contrast, to prevent the transmission of data frames during WPT, information on the timings of WPT and data transmission must be shared.

Note that some results presented in this chapter are specific to WLAN devices used in this study (e.g., rate adaptation and adjacent channel rejection). However, we want to emphasize that the purpose of this chapter was to reveal unknown issues that are caused by applying WPT to WLAN devices.

Acknowledgment

We greatly thank Prof. Masahiro Morikura, Prof. Naoki Shinohara, Dr. Takayuki Nishio, Mr. Norikatsu Imoto, Mr. Koichi Sakaguchi, Mr. Takuya Ichihara, and Dr. Yong Huang for their supports. This experiment was carried out using the Microwave Energy Transmission Laboratory (METLAB) system of the Research Institute for Sustainable Humanosphere at Kyoto University.

References

[1] ITU-R. (2016). *Applications of Wireless Power Transmission via Radio Frequency Beam*. Technical Report, ITU-R SM.2392-0. Geneva: The International Telecommunication Union.

[2] Imoto, N., Yamashita, S., Ichihara, T., Yamamoto, K., Nishio, T., Morikura, M., et al. (1842). Experimental investigation of co-channel and adjacent channel operations of microwave power and IEEE 802.11g data transmissions. *IEICE Trans. Commun.* E97-B, 1835–1842.

[3] Yamashita, S., Imoto, N., Ichihara, T., Yamamoto, K., Nishio, T., Morikura, M., et al. (2014). Implementation and feasibility study of co-channel operation system of microwave power transmissions to IEEE 802.11-based batteryless sensor. *IEICE Trans. Commun.*, E97-B, 1843–1852.

[4] Yamashita, S., Sakaguchi, K., Huang, Y., Yamamoto, K., Nishio, T., Morikura, M., et al. (2015). Rate adaptation based on exposure assessment using rectenna output for WLAN station powered with microwave power transmission. *IEICE Trans. Commun.* E98-B, 1785–1794.

[5] Umeda, T., Yoshida, H., Sekine, S., Fujita, Y., Suzuki, T., and Otaka, S. (2006). A 950-MHz rectifier circuit for sensor network tags with 10-m distance. *IEEE J. Solid State Circuits* 41, 35–41.

[6] Yoshida, S., Noji, T., Fukuda, G., Kobayashi, Y., and Kawasaki, S. (2013). Experimental demonstration of coexistence of microwave wireless communication and power transfer technologies for battery-free sensor network systems. *Int. J. Anntenas Propag.* 1–10.

[7] Shinohara, N., Tomohiko, M., and Matsumoto, H. (2005). "Study on ubiquitous power source with microwave power transmission," in *Proceedings of the Union Radio Science (URSI) General Assembly 2005*, New Delhi, 1–4.

[8] Farinholt, K. M., Park, G., and Farrar, C. R. (2009). RF energy transmission for a low-power wireless impedance sensor node. *IEEE Sens. J.* 9, 793–800.

[9] Ichihara, T., Mitani, T., and Shinohara, N. (2012). "Study on intermittent microwave power transmission to a ZigBee device," in *Proceedings of the IEEE Microwave Workshop Series (IMWS) on Innovative Wireless Power Transmission: Technologies, Systems, and Applications 2012*, Kyoto, 209–212.

[10] Brown, W. C. (1984). The history of power transmission by radio waves. *IEEE Trans. Microw. Theory Technol.* 32, 1230–1242.

[11] Paing, T., Morroni, J., Dolgov, A., Shin, J., Brannan, J., Zane, R., et al. (2007). "Wirelessly-powered wireless sensor platform," in *Proceedings of the European Microwave Conference 2007*, Munich, 241–244.

[12] Iperf. Available at: http://www. iperf. fr

[13] Shacham, N., and McKenney, P. (1990). "Packet recovery in high-speed networks using coding and buffer management," in *Proceedings of the IEEE International Conference on Computer Communications (INFOCOM)*, San Francisco, CA, 124–131.

[14] Zhang, J., Tan, K., Zhao, J., Wu, H., and Zhang, Y. (2008). "A practical SNR-guided rate adaptation," in *Proceedings of the IEEE International Conference Computer Communication (INFOCOM)*, Phoenix, AZ, 146–150.

[15] Pelechrinis, K., Broustis, I., Krishnamurthy, S. V., and Gkantsidis, C.-T. (2011). A measurement-driven anti-jamming system for 802.11 networks. *IEEE/ACM Trans. Netw.* 19, 1208–1222.

[16] Aguayo, D., Bicket, J., Biswas, S., Judd, G., and Morris, R. (2004). "Link-level measurements from an 802.11b mesh network," in *Proceedings of the ACM Annual Conference Special Interest Group Data Commun (SIGCOMM)*, New York, 121–132.

Index

About the Editor

Naoki Shinohara received the B.E. degree in electronic engineering, the M.E. and Ph.D. (Eng.) degrees in electrical engineering from Kyoto University, Japan, in 1991, 1993 and 1996, respectively. He was a research associate in Kyoto University from 1996. From 2010, he has been a professor in Kyoto University. He has been engaged in research on Solar Power Station/Satellite and Microwave Power Transmission system. He is IEEE MTT-S Technical Committee 26 (Wireless Power Transfer and Conversion) chair, IEEE MTT-S Kansai Chapter TPC member, IEEE Wireless Power Transfer Conference advisory committee member, URSI Commission D vice chair, international journal of Wireless Power Transfer (Cambridge Press) executive editor, first chair and technical committee member on IEICE Wireless Power Transfer, Japan Society of Electromagnetic Wave Energy Applications vice president, Space Solar Power Systems Society board member, Wireless Power Transfer Consortium for Practical Applications (WiPoT) chair, and Wireless Power Management Consortium (WPMc) chair.